EXTRAORDINARY ACCLAIM FOR JONAH LEHRER

AND *Proust Was a Neuroscientist*

"A precocious and engaging book that tries to mend the century-old tear between the literary and scientific cultures . . . Lehrer is smart, and there are some fun moments in these pages." — *New York Times Book Review*

"A slim, brainy book about the brain, modernist art, and literature . . . Lehrer writes skillfully and coherently about both art and science — no small feat." — *Entertainment Weekly*

"Lehrer is most surprising and inventive when correlating how Stravinsky's 'Rite of Spring' changed the way we hear music with experiments that later tracked neurons in the auditory cortex and how Escoffier's innovations in the kitchen were designed to stimulate specific receptors that were later located on the tongue." — *Boston Globe*

"Entertaining and enlightening, and Lehrer gets extra points for a mouthwatering section on Escoffier's experiments with sauces." — *New York*

"Lehrer is gifted with the ability to find philosophy in science and stray bits of science buried amid the rubble of literary history. He is less critic than armchair philosopher, searching for meaning anywhere great thinkers have left their footprints." — *San Francisco Chronicle*

"A remarkable, fun, and intriguing debut: a knowing work of art about art's meeting with the science of knowing." — *New York Post*

"Lehrer's real aim is to demonstrate that the words and sounds of these creative souls revealed how our brains not only record the world but help to create it in the act of perception." — *Dallas Morning News*

"Jonah Lehrer's smart, elegantly written little book expresses an appealing faith that art and science offer different but complementary views of the world." — *Washington Post Book World*

"A captivating, fast-paced read that should appeal to scientists and art lovers alike." — *Chicago Sun-Times*

Proust Was a Neuroscientist

JONAH LEHRER

A MARINER BOOK
HOUGHTON MIFFLIN COMPANY
Boston · New York

For Sarah and Ariella

First Mariner Books edition 2008
Copyright © 2007 by Jonah Lehrer
ALL RIGHTS RESERVED

www.hmhco.com

Library of Congress Cataloging-in-Publication Data
Lehrer, Jonah.
Proust was a neuroscientist / Jonah Lehrer.
p. cm.
Includes bibliographical references and index.
ISBN-13: 978-0-618-62010-4
ISBN-10: 0-618-62010-9
1. Neurosciences and the arts.
2. Neurosciences — History. I. Title.
NX180.N48L44 2007 700.1'05 — dc22 2007008518

ISBN 978-0-547-08590-6 (pbk.)

Printed in the United States of America

Book design by Robert Overholtzer

DOC 10

PHOTO CREDITS: page 6, Courtesy of the Oscar Lion Collection, Rare
Books Division, The New York Public Library, Astor, Lenox and Tilden
Foundations; page 23, Courtesy of the Henry W. and Albert A. Berg
Collection of English and American Literature, The New York Public
Library, Astor, Lenox, and Tilden Foundations; page 30, Courtesy of
Image Select/Art Resource, New York; page 77, Courtesy Erich Lessing/Art
Resource, New York; page 86, Courtesy of Christie's Images; page 103,
Courtesy of Musée d'Orsay/Erich Lessing/Art Resource, New York;
page 115, Courtesy of Galerie Beyeler/Bridgeman-Giraudon/Art Resource,
New York; page 124, Courtesy Leopold Stokowski Collection of Conducting
Scores, Rare Book & Manuscript Library, University of Pennsylvania;
page 150, Courtesy of the Metropolitan Museum of Art (47.106);
page 170, Courtesy of the New York Public Library/Art Resource, New
York; page 174, Courtesy British Library/HIP/Art Resource, New York

Contents

Reality is a product of the most august imagination.

—Wallace Stevens

This systematic denial on science's part of personality
as a condition of events, this rigorous belief that in its
own essential and innermost nature our world is a strictly
impersonal world, may, conceivably, as the whirligig of
time goes round, prove to be the very defect that our
descendants will be most surprised at in our own boasted
science, the omission that to their eyes will most tend
to make it look perspectiveless and short.

—William James

Prelude

I used to work in a neuroscience lab. We were trying to figure out how the mind remembers, how a collection of cells can encapsulate our past. I was just a lab technician, and most of my day was spent performing the strange verbs of bench science: amplifying, vortexing, pipetting, sequencing, digesting, and so on. It was simple manual labor, but the work felt profound. Mysteries were distilled into minor questions, and if my experiments didn't fail, I ended up with an answer. The truth seemed to slowly accumulate, like dust.

At the same time, I began reading Proust. I would often bring my copy of *Swann's Way* into the lab and read a few pages while waiting for an experiment to finish. All I expected from Proust was a little entertainment, or perhaps an education in the art of constructing sentences. For me, his story about one man's memory was simply that: a story. It was a work of fiction, the opposite of scientific fact.

But once I got past the jarring contrast of forms—my science spoke in acronyms, while Proust preferred meandering prose—I began to see a surprising convergence. The novelist had predicted my experiments. Proust and neuroscience shared a vision of how our memory works. If you listened closely, they were actually saying the same thing.

This book is about artists who anticipated the discoveries of neuroscience. It is about writers and painters and composers who discovered truths about the human mind—real, tangible truths—that science is only now *re*discovering. Their imaginations foretold the facts of the future.

Of course, this isn't the way knowledge is supposed to advance. Artists weave us pretty tales, while scientists objectively describe the universe. In the impenetrable prose of the scientific paper, we imagine a perfect reflection of reality. One day, we assume, science will solve everything.

In this book, I try to tell a different story. Although these artists witnessed the birth of modern science—Whitman and Eliot contemplated Darwin, Proust and Woolf admired Einstein—they never stopped believing in the necessity of art. As scientists were beginning to separate thoughts into their anatomical parts, these artists wanted to understand consciousness from the inside. Our truth, they said, must begin with us, with what reality *feels* like.

Each of these artists had a peculiar method. Marcel Proust spent all day in bed, ruminating on his past. Paul Cézanne would stare at an apple for hours. Auguste Escoffier was just trying to please his customers. Igor Stravinsky was trying *not* to please his customers. Gertrude Stein liked to play with words. But despite their technical differences, all of these artists shared an abiding interest in human experience. Their creations were acts of exploration, ways of grappling with the mysteries they couldn't understand.

These artists lived in an age of anxiety. By the middle of the nineteenth century, as technology usurped romanticism, the essence of human nature was being questioned. Thanks to the distressing discoveries of science, the immortal soul was dead. Man was a monkey, not a fallen angel. In the frantic search for new kinds of expression, artists came up with a new method: they looked in the mirror. (As Ralph Waldo Emerson declared, "The mind has become aware of itself.") This inward turn created art that was exquisitely self-conscious; its subject was our psychology.

The birth of modern art was messy. The public wasn't accustomed to free-verse poems or abstract paintings or plotless novels. Art was supposed to be pretty or entertaining, preferably both. It was supposed to tell us stories about the world, to give us life as it should be, or could be. Reality was hard, and art was our escape. But the modernists refused to give us what we wanted. In a move of stunning arrogance and ambition, they tried to invent fictions that told the truth. Although their art was difficult, they aspired to transparency: in the forms and fractures of their work, they wanted us to see ourselves.

The eight artists in this book are not the only people who tried to understand the mind. I have chosen them because their art proved to be the most accurate, because they most explicitly anticipated our science. Nevertheless, the originality of these artists was influenced by a diverse range of other thinkers. Whitman was inspired by Emerson, Proust imbibed Bergson, Cézanne studied Pissarro, and Woolf was emboldened by Joyce. I have attempted to sketch the intellectual atmosphere that shaped their creative process, to highlight the people and ideas from which their art emerged.

One of the most important influences on all of these artists—and the only influence they all shared—was the science of their time. Long before C. P. Snow mourned the sad separation of our two cultures, Whitman was busy studying brain anatomy textbooks and watching gruesome surgeries, George Eliot was reading Darwin and James Clerk Maxwell, Stein was conducting psychology experiments in William James's lab, and Woolf was learning about the biology of mental illness. It is impossible to understand their art without taking into account its relationship to science.

This was a thrilling time to be studying science. By the start of the twentieth century, the old dream of the Enlightenment seemed within reach. Everywhere scientists looked, mystery seemed to retreat. Life was just chemistry, and chemistry was just physics. The entire universe was nothing but a mass of vibrating molecules. For the most part, this new knowledge represented the triumph of a method; scientists had discovered reductionism and were successfully applying it to reality. In Plato's metaphor, the reductionist aims to "cut nature at its joints, like a good butcher." The whole can be understood only by breaking it apart, dissecting reality until it dissolves. This is all we are: parts, acronyms, atoms.

But these artists didn't simply translate the facts of science into pretty new forms. That would have been too easy. By exploring their own experiences, they expressed what no experiment could see. Since then, new scientific theories have come and gone, but this art endures, as wise and resonant as ever.

We now know that Proust was right about memory, Cézanne was uncannily accurate about the visual cortex, Stein anticipated Chomsky, and Woolf pierced the mystery of consciousness; modern neuroscience has confirmed these artistic intuitions. In each of the following chapters, I have tried to give a sense of the scientific process, of how scientists actually distill their data into rigorous new hypotheses. Every brilliant experiment, like every great work of art, starts with an act of imagination.

Unfortunately, our current culture subscribes to a very narrow definition of truth. If something can't be quantified or calculated, then it can't be true. Because this strict scientific approach has explained so much, we assume that it can explain everything. But every method, even the experimental method, has limits. Take the human mind. Scientists describe our brain in terms of its physical details; they say we are nothing but a loom of electrical cells and synaptic spaces. What science forgets is that this isn't how we experience the world. (We feel like the ghost, not like the machine.) It is ironic but true: the one reality science cannot reduce is the only reality we will ever know. This is why we need art. By expressing our actual experience, the artist reminds us that our science is incomplete, that no map of matter will ever explain the immateriality of our consciousness.

The moral of this book is that we are made of art and science. We are such stuff as dreams are made on, but we are also just stuff. We now know enough about the brain to realize that its mystery will always remain. Like a work of art, we exceed our materials. Science needs art to frame the mystery, but art needs science so that not everything is a mystery. Neither truth alone is our solution, for our reality exists in plural.

I hope these stories of artistic discovery demonstrate that any description of the brain requires both cultures, art and science. The reductionist methods of science must be allied with an artistic investigation of our experience. In the following chapters, I try to re-imagine this dialogue. Science is seen through the optic of art, and art is interpreted in the light of science. The experiment and the poem complete each other. The mind is made whole.

*Proust Was
a Neuroscientist*

Chapter 1

Walt Whitman

The Substance of Feeling

> The poet writes the history of his own body.
>
> — Henry David Thoreau

FOR WALT WHITMAN, the Civil War was about the body. The crime of the Confederacy, Whitman believed, was treating blacks as nothing but flesh, selling them and buying them like pieces of meat. Whitman's revelation, which he had for the first time at a New Orleans slave auction, was that body and mind are inseparable. To whip a man's body was to whip a man's soul.

This is Whitman's central poetic idea. We do not *have* a body, we *are* a body. Although our feelings feel immaterial, they actually begin in the flesh. Whitman introduces his only book of poems, *Leaves of Grass,* by imbuing his skin with his spirit, "the aroma of my armpits finer than prayer":

> Was somebody asking to see the soul?
> See, your own shape and countenance . . .
> Behold, the body includes and is the meaning, the main
> Concern, and includes and is the soul

Whitman's fusion of body and soul was a revolutionary idea, as radical in concept as his free-verse form. At the time, scientists believed that our feelings came from the brain and that the body was just a lump of inert matter. But Whitman believed that our mind depended upon the flesh. He was determined to write poems about our "form complete."

This is what makes his poetry so urgent: the attempt to wring "beauty out of sweat," the metaphysical soul out of fat and skin. Instead of dividing the world into dualisms, as philosophers had done for centuries, Whitman saw everything as continuous with everything else. For him, the body and the soul, the profane and the profound, were only different names for the same thing. As Ralph Waldo Emerson, the Boston Transcendentalist, once declared, "Whitman is a remarkable mixture of the Bhagvat Ghita and the *New York Herald*."

Whitman got this theory of bodily feelings from his investigations of himself. All Whitman wanted to do in *Leaves of Grass* was put "a *person*, a human being (myself, in the later half of the Nineteenth Century, in America) freely, fully and truly on record." And so the poet turned himself into an empiricist, a lyricist of his own experience. As Whitman wrote in the preface to *Leaves of Grass*, "You shall stand by my side to look in the mirror with me."

It was this method that led Whitman to see the soul and body as inextricably "interwetted." He was the first poet to write poems in which the flesh was not a stranger. Instead, in Whitman's unmetered form, the landscape of his body became the inspiration for his poetry. Every line he ever wrote ached with the urges of his anatomy, with its wise desires and inarticulate sympathies. Ashamed of nothing, Whitman left nothing out. "Your very flesh," he promised his readers, "shall be a great poem."

Neuroscience now knows that Whitman's poetry spoke the truth: emotions are generated by the body. Ephemeral as they seem, our feelings are actually rooted in the movements of our muscles and the palpitations of our insides. Furthermore, these material feelings are an essential element of the thinking process. As the neuroscientist Antonio Damasio notes, "The mind is embodied . . . not just embrained."

At the time, however, Whitman's idea was seen as both erotic and audacious. His poetry was denounced as a "pornographic utterance," and concerned citizens called for its censorship. Whitman enjoyed the controversy. Nothing pleased him more than dismantling prissy Victorian mores and inverting the known facts of science.

The story of the brain's separation from the body begins with René Descartes. The most influential philosopher of the seventeenth century, Descartes divided being into two distinct substances: a holy soul and a mortal carcass. The soul was the source of reason, science, and everything nice. Our flesh, on the other hand, was "clocklike," just a machine that bleeds. With this schism, Descartes condemned the body to a life of subservience, a power plant for the brain's light bulbs.

In Whitman's own time, the Cartesian impulse to worship the brain and ignore the body gave rise to the new "science" of phrenology. Begun by Franz Josef Gall at the start of the nineteenth century, phrenologists believed that the shape of the skull, its strange hills and hollows, accurately reflected the mind inside. By measuring the bumps of bone, these pseudoscientists hoped to measure the subject's character by determining which areas of the brain were swollen with use and which were shriveled with neglect. Our cranial packaging revealed our insides; the rest of the body was irrelevant.

By the middle of the nineteenth century, the promise of phrenology seemed about to be fulfilled. Innumerable medical treatises, dense with technical illustrations, were written to defend its theories. Endless numbers of skulls were quantified. Twenty-seven different mental talents were uncovered. The first scientific theory of mind seemed destined to be the last.

But measurement is always imperfect, and explanations are easy to invent. Phrenology's evidence, though amassed in a spirit of seriousness and sincerity, was actually a collection of accidental observations. (The brain is so complicated an organ that its fissures can justify almost any imaginative hypothesis, at least until a better hypothesis comes along.) For example, Gall located the trait of ideality in "the temporal ridge of the frontal bones" because busts of Homer revealed a swelling there and because poets when writing tend to touch that part of the head. This was his data.

Of course, phrenology strikes our modern sensibilities as woefully unscientific, like an astrology of the brain. It is hard to imagine its allure or comprehend how it endured for most of the nineteenth century.* Whitman used to quote Oliver Wendell Holmes on the subject: "You might as easily tell how much money is in a safe feeling the knob on the door as tell how much brain a man has by feeling the bumps on his head." But knowledge emerges from the litter of our mistakes, and just as alchemy led to chemistry, so did the failure of phrenology lead science to study the brain itself and not just its calcified casing.

Whitman, a devoted student of the science of his day,† had a complicated relationship with phrenology. He called the first phrenology lecture he attended "the greatest conglomeration of pretension and absurdity it has *ever* been our lot to listen to. . . . We do not mean to assert that there is no truth whatsoever in phrenology, but we do say that its claims to confidence, as set forth by Mr. Fowler, are preposterous to the last degree." More than a decade later, however, that same Mr. Fowler, of the publishing house Fowler and Wells in Manhattan, became the sole distributor of the first edition of *Leaves of Grass*. Whitman couldn't find anyone else to publish his poems. And while Whitman seems to have moderated his views on

---

* The single biggest failing of phrenology was its inability to assimilate data that didn't conform to its predictions. For example, when phrenologists measured Descartes's skull, they found an extremely small forehead, which implied "limited logical and rational faculties." But instead of doubting their original hypothesis, the phrenologists lampooned Descartes and declared "that he was not so great a thinker as he was held to be."

† Although Whitman loved learning about science, he never accepted its findings uncritically. In his notebook, Whitman reminded himself to always question the veracity of its experiment: "Remember in scientific and similar allusions that the theories of Geology, History, Language, &c., &c., are continually changing. Be careful to put in only what must be appropriate centuries hence."

the foolishness of phrenology — even going so far as to undergo a few phrenological exams himself* — his poetry stubbornly denied phrenology's most basic premise. Like Descartes, phrenologists looked for the soul solely in the head, desperate to reduce the mind to its cranial causes. Whitman realized that such reductions were based on a stark error. By ignoring the subtleties of his body, these scientists could not possibly account for the subtleties of his soul. Like *Leaves of Grass,* which could only be understood in "its totality — its massings," Whitman believed that his existence could be "comprehended at no time by its parts, at all times by its unity." This is the moral of Whitman's poetic sprawl: the human being is an irreducible whole. Body and soul are emulsified into each other. "To be in any form, what is that?" Whitman once asked. "Mine is no callous shell."

Emerson

Whitman's faith in the transcendental body was strongly influenced by the transcendentalism of Ralph Waldo Emerson. When Whitman was still a struggling journalist living in Brooklyn, Emerson was beginning to write his lectures on nature. A lapsed Unitarian preacher, Emerson was more interested in the mystery of his own mind than in the preachings of some aloof God. He disliked organized religion because it relegated the spiritual to a place in the sky instead of seeing the spirit among "the common, low and familiar."

Without Emerson's mysticism, it is hard to imagine Whitman's poetry. "I was simmering, simmering, simmering," Whitman once said, "and Emerson brought me to a boil." From Emerson, Whitman learned to trust his own experience, searching himself for intimations of the profound. But if the magnificence of Emerson was

---

* Whitman was not alone; everyone from Mark Twain to Edgar Allan Poe underwent phrenological exams. George Eliot shaved her head so that her phrenologist could make a more accurate diagnosis of her cranial bumps.

*An engraving of Walt Whitman from July 1854. This image
served as the frontispiece for the first edition of* Leaves of Grass.

his vagueness, his defense of Nature with a capital *N*, the magnificence of Whitman was his immediacy. All of Whitman's songs began with himself, nature as embodied by his own body.

And while Whitman and Emerson shared a philosophy, they
could not have been more different in person. Emerson looked like
a Puritan minister, with abrupt cheekbones and a long, bony nose.
A man of solitude, he was prone to bouts of selfless self-absorption.

"I like the silent church before the service begins," he confessed in "Self-Reliance." He wrote in his journal that he liked man, but not men. When he wanted to think, he would take long walks by himself in the woods.

Whitman — "broad shouldered, rough-fleshed, Bacchus-browed, bearded like a satyr, and rank" — got his religion from Brooklyn, from its dusty streets and its cart drivers, its sea and its sailors, its mothers and its men. He was fascinated by people, these citizens of his sensual democracy. As his uncannily accurate phrenological exam put it,* "Leading traits of character appear to be Friendship, Sympathy, Sublimity and Self-Esteem, and markedly among his combinations the dangerous fault of Indolence, a tendency to the pleasure of Voluptuousness and Alimentiveness, and a certain reckless swing of animal will, too unmindful, probably, of the conviction of others."

Whitman heard Emerson for the first time in 1842. Emerson was beginning his lecture tour, trying to promote his newly published *Essays.* Writing in the New York *Aurora*, Whitman called Emerson's speech "one of the richest and most beautiful compositions" he had ever heard. Whitman was particularly entranced by Emerson's plea for a new American poet, a versifier fit for democracy: "The poet stands among partial men for the complete man," Emerson said. "He reattaches things to the whole."

But Whitman wasn't ready to become a poet. For the next decade, he continued to simmer, seeing New York as a journalist and as the editor of the *Brooklyn Eagle* and *Freeman*. He wrote articles about criminals and abolitionists, opera stars and the new Fulton ferry. When the *Freeman* folded, he traveled to New Orleans, where he saw slaves being sold on the auction block, "their bodies encased in metal chains." He sailed up the Mississippi on a side-wheeler, and

---

* One of the reasons that Whitman seems to have moderated his views on phrenology was that he had a very favorable skull. He scored nearly perfect on virtually every possible phrenological trait. Oddly enough, two of his lowest scores were for the traits of tune and language.

got a sense of the Western vastness, the way the "United States themselves are essentially the greatest poem."

It was during these difficult years when Whitman was an unemployed reporter that he first began writing fragments of poetry, scribbling down quatrains and rhymes in his cheap notebooks. With no audience but himself, Whitman was free to experiment. While every other poet was still counting syllables, Whitman was writing lines that were messy montages of present participles, body parts, and erotic metaphors. He abandoned strict meter, for he wanted his form to reflect nature, to express thoughts "so alive that they have an architecture of their own." As Emerson had insisted years before, "Doubt not, O poet, but persist. Say 'It is in me, and shall out.'"

And so, as his country was slowly breaking apart, Whitman invented a new poetics, a form of inexplicable strangeness. A self-conscious "language-maker," Whitman had no precursor. No other poet in the history of the English language prepared readers for Whitman's eccentric cadences ("sheath'd hooded sharp-tooth'd touch"), his invented verbs ("unloosing," "preluding," "unreeling"), his love of long anatomical lists,* and his honest refusal to be anything but himself, syllables be damned. Even his bad poetry is bad in a completely original way, for Whitman only ever imitated himself.

And yet, for all its incomprehensible originality, Whitman's verse also bears the scars of his time. His love of political unions and physical unity, the holding together of antimonies: these themes find their source in America's inexorable slide into the Civil War. "My book and the war are one," Whitman once said. His notebook breaks into free verse for the first time in lines that try to unite the decade's irreconcilables, the antagonisms of North and South, master and slave, body and soul. Only in his poetry could Whitman find the whole he was so desperately looking for:

---

* "The lung sponges, the stomach-sac, the bowels sweet and clean,
  The brain in its folds inside the skull frame,
  Sympathies, heart-valves, palate valves, sexuality, maternity . . ." [129]

> I am the poet of the body
> And I am the poet of the soul
> I go with the slaves of the earth equally with the masters
> And I will stand between the masters and the slaves,
> Entering into both so that both shall understand me alike.

In 1855, after years of "idle versifying," Whitman finally published his poetry. He collected his "leaves" — printing lingo for pages — of "grass" — what printers called compositions of little value — in a slim, cloth-bound volume, only ninety-five pages long. Whitman sent Emerson the first edition of his book. Emerson responded with a letter that some said Whitman carried around Brooklyn in his pocket for the rest of the summer. At the time, Whitman was an anonymous poet and Emerson a famous philosopher. His letter to Whitman is one of the most generous pieces of praise in the history of American literature. "Dear Sir," Emerson began:

> I am not blind to the worth of the wonderful gift of "Leaves of Grass." I find it the most extraordinary piece of wit & wisdom that America has yet contributed. I am very happy in reading it. It meets the demand I am always making of what seemed the sterile & stingy nature, as if too much handiwork or too much lymph in the temperament were making our western wits fat & mean. I give you joy of your free & brave thought. . . . I greet you at the beginning of a great career.

Whitman, never one to hide a good review from "the Master," sent Emerson's private letter to the *Tribune,* where it was published and later included in the second edition of *Leaves of Grass.* But by 1860, Emerson had probably come to regret his literary endorsement. Whitman had added to *Leaves of Grass* the erotic sequence "Enfans d'Adam" ("Children of Adam"), a collection that included the poems "From Pent-up Aching Rivers," "I Am He that Aches with Love," and "O Hymen! O Hymenee!" Emerson wanted Whitman to remove the erotic poems from the new edition of his poetry. (Apparently, some parts of Nature still had to be censored.) Emerson made this clear while the two were taking a long walk across

Boston Common, expressing his fear that Whitman was "in danger of being tangled up with the unfortunate heresy" of free love.

Whitman, though still an obscure poet, was adamant: "Enfans d'Adam" must remain. Such an excision, he said, would be like castration and "What does a man come to with his virility gone?" For Whitman, sex revealed the unity of our form, how the urges of the flesh became the feelings of the soul. He would remember in the last preface to *Leaves of Grass*, "A Backwards Glance over Traveled Roads," that his conversation with Emerson had crystallized his poetic themes. Although he admitted that his poetry was "avowedly the song of sex and Amativeness and ever animality," he believed that his art "lifted [these bodily allusions] into a different light and atmosphere." Science and religion might see the body in terms of its shameful parts, but the poet, lover of the whole, knows that "the human body and soul must remain an entirety." "*That*," insisted Whitman, "is what I felt in my inmost brain and heart, when I only answer'd Emerson's vehement arguments with silence, under the old elms of Boston Common."

Despite his erotic epiphany, Whitman was upset by his walk with Emerson. Had no one understood his earlier poetry? Had no one seen its philosophy? *The body is the soul.* How many times had he written that? In how many different ways? And if the body is the soul, then how can the body be censored? As he wrote in "I Sing the Body Electric," the central poem of "Enfans d'Adam":

> O my body! I dare not desert the likes of you in other men
> and women, nor the likes of the parts of you,
> I believe the likes of you are to stand or fall with the likes
> of the soul, (and that they are the soul,)
> I believe the likes of you shall stand or fall with my
> Poems, and that they are my poems.

And so, against Emerson's wishes, Whitman published "Enfans d'Adam." As Emerson predicted, the poems were greeted with cries of indignation. One reviewer said "that quotations from the 'Enfans d'Adam' poems would be an offence against decency too gross to be tolerated." But Whitman didn't care. As usual, he wrote his own anonymous reviews. He knew that if his poetry were to last, it must leave nothing out. It must be candid, and it must be true.

The Ghostly Limb

In the winter of 1862, during the bloody apogee of the Civil War, Whitman traveled to Virginia in search of his brother, who had been injured at the Battle of Fredericksburg. This was Whitman's first visit to the war's front. The fighting had ended just a few days before, and Whitman saw "where their priceless blood reddens the grass the ground." The acrid smell of gun smoke still hung in the air. Eventually, Whitman found the Union Army hospital, its shelter tents bordered by freshly dug graves, the names of the dead scrawled on "pieces of barrel-staves or broken boards, stuck in the dirt." Writing to his mother, Whitman described "the heap of feet, arms, legs &c. under a tree in front of a hospital." The limbs, freshly amputated, were beginning to rot.

After seeing the dead and dying of Fredericksburg, Whitman devoted himself to helping the soldiers. For the next three years, he volunteered as a wound dresser in Union hospitals, seeing "some 80,000 to 100,000 of the wounded and sick, as sustainer of spirit and body in some degree." He would nurse both Union and Confederate men. "I cannot leave them," he wrote. "Once in a while some youngster holds on to me convulsively and I do what I can for him." Whitman held the soldiers' hands; he made them lemonade; he bought them ice cream and underwear and cigarettes; sometimes, he even read them poetry. While the doctors treated their wounds, Whitman nursed their souls.

All his life, Whitman would remember the time he spent as a volunteer in the hospitals. "Those three [wartime] years," he later remembered in *Specimen Days,* his oral autobiography, "I consider the most profound lesson of my life." Never again would Whitman feel so useful, "more permanently absorbed, to the very roots." "People used to say to me, 'Walt, you are doing miracles for those fellows in the hospitals.' I wasn't. I was . . . doing miracles for my-self."

As always, Whitman transmuted the experience into poetry. He told Emerson that he wanted to write about his time in the hospitals, for they had "opened a new world somehow to me, giving closer insights, new things, exploring deeper mines than any yet." In "Drum Taps," his sequence of poems on the war — the only sequence of poems he never revised — Whitman describes the tortured anatomy he saw every day in the hospitals:

> From the stump of the arm, the amputated hand
> I undo the clotted lint, remove the slough, wash off the
> matter and blood,
> Back on his pillow the soldier bends with curv'd neck and
> side-falling head,
> His eyes are closed, his face pale, he dares not look on the
> bloody stump.

Whitman did look at the bloody stump. The war's gore shocked him. Volunteering in the canvas-tent hospitals, he witnessed the violent mess of surgery: "the hiss of the surgeon's knife, the gnawing teeth of his saw / wheeze, cluck, swash of falling blood." Amid the stench of dying soldiers and unclaimed corpses, Whitman consoled himself by remembering that the body was not only a body. As a nurse, Whitman tried to heal what the surgeon couldn't touch. He called these our "deepest remains."

By the second year of the war, just as Whitman was learning how to wrap battle wounds in wet cotton, doctors working in Civil War hospitals began noticing a very strange phenomenon. After a soldier's limb was amputated, it was not uncommon for him to continue to "feel" his missing arm or leg. The patients said it was like living with ghosts. Their own flesh had returned to haunt them.

Medical science ignored the syndrome. After all, the limb and its nerves were gone. There was nothing left to cut. But one doctor believed the soldiers' strange stories. His name was Silas Weir Mitchell, and he was a "doctor of nerves" at Turner's Lane Hospital in Philadelphia. He was also a good friend of Whitman's. For much of their lives, the doctor and the poet wrote letters to each other, sharing a love of literature and medical stories. In fact, it was Weir Mitchell who, in 1878, finally diagnosed Whitman with a ruptured blood vessel in the brain, prescribing "mountain air" as medicine. Later on, Weir Mitchell financially supported the poet, giving him fifteen dollars a month for more than two years.

But during the Civil War, while Whitman was working as a nurse, Weir Mitchell was trying to understand these illusory limbs. The Battle of Gettysburg had given him a hospital full of amputee patients, and, in his medical notebook, Weir Mitchell began describing a great variety of "sensory ghosts." Some of the missing limbs seemed unreal to the patients, while others seemed authentic; some were painful, others painless. Although a few of the amputees eventually forgot about their amputated limbs, the vast majority retained "a sense of the existence of their lost limb that was more vivid, definite and intrusive than that of its truly living fellow member." The bodily illusion was more real than the body.

Although Weir Mitchell believed that he was the first person to document this phenomenon, he wasn't. Herman Melville, twelve years earlier, had given Ahab, the gnarled sea captain of *Moby-Dick,* a sensory ghost. Ahab is missing a leg (Moby-Dick ate it), and in chapter 108, he summons a carpenter to fashion him a new ivory peg leg. Ahab tells the carpenter that he still feels his amputated leg "invisibly and uninterpenetratingly." His phantom limb is like a "poser." "Look," Ahab says, "put thy live leg here in the place where mine was; so, now, here is only one distinct leg to the eye, yet two to the soul. Where thou feelest tingling life; there, exactly there, there to a hair, do I. Is't a riddle?"

Weir Mitchell, unaware of Melville's prescience, never cited Ahab's medical condition. He published his observations of the mystery in two neurology textbooks. He even published a special bulletin on the phenomenon, which the surgeon general's office distributed to other military hospitals in 1864. But Weir Mitchell felt constrained by the dry, clinical language of his medical reports. He believed that the experience of the soldiers in his hospital had profound philosophical implications. After all, their sensory ghosts were living proof of Whitman's poetry: our matter was entangled with our spirit. When you cut the flesh, you also cut the soul.

And so Weir Mitchell decided to write an anonymous short story, written in the first person.* In "The Case of George Dedlow," published in *The Atlantic Monthly* in 1866, Weir Mitchell imagines himself a soldier wounded at the Battle of Chickamauga, shot in both legs and both arms. Dedlow passes out from the pain.

When he wakes, Dedlow is in a hospital tent. He has no limbs left: they have all been cut off. Dedlow describes himself as a "useless torso, more like some strange larval creature than anything of human shape." But even though Dedlow is now limbless, he still *feels* all of his limbs. His body has become a ghost, and yet it feels as real as ever. Weir Mitchell explains this phenomenon by referencing the brain. Because the brain and body are so interconnected, the mind remains "ever mindful of its missing [bodily] part, and, imperfectly at least, preserves to the man a consciousness of possessing that which he has not." Weir Mitchell believed that the brain depended upon the body for its feelings and identity. Once Dedlow loses his limbs, he finds "to his horror that at times I was less conscious of myself, of my own existence, than used to be the case . . . I thus reached the conclusion that *a man is not his brain, or any one part of it, but all of his economy,* and that to lose any part must lessen this sense of his own existence."

In his short story, Weir Mitchell is imagining a Whitmanesque physiology. Since soul is body and body is soul, to lose a part of one's body is to lose a part of one's soul. As Whitman wrote in "Song of Myself," "Lack one lacks both." The mind *cannot* be extricated from its matter, for mind and matter, these two seemingly op-

---

* Later on in his life, Weir Mitchell would abandon medicine entirely and devote himself to writing novels and poetry. His novel *Hugh Wynne* (1897) — about the experiences of a Quaker during the American Revolution — was particularly popular.

posite substances, are impossibly intertwined. Whitman makes our unity clear on the very first page of *Leaves of Grass,* as he describes his poetic subject:

> Of physiology from top to toe I sing
> not physiognomy alone nor brain alone is worthy for the
> Muse, I say the form complete is worthier far.

After the war, Weir Mitchell's clinical observations fell into obscurity. Because phantom limbs had no material explanation, medical science continued to ignore the phenomenon. Only William James, in his 1887 article "The Consciousness of Lost Limbs," pursued Weir Mitchell's supernatural hypothesis.* As Harvard's first psychology professor, James sent out a short questionnaire to hundreds of amputees asking various questions about their missing parts (for example, "How much of the limb can you feel?" "Can you, by *imagining* strongly that it has moved, make yourself really feel as if it *had* moved into a different position?"). The results of James's survey taught him only one fact about sensory ghosts: there was no general pattern to the experience of lost limbs. Every body was invested with its own individual meaning. "We can never seek amongst these processes for results which shall be invariable," James wrote. "Exceptions remain to every empirical law of our mental life, and can only be treated as so many individual aberrations." As Henry James, William's novelist brother, once wrote, "There is a presence in what is missing." That presence is our own.

The Anatomy of Emotion

Whitman's faith in the flesh, although it was the source of his censorship, had a profound impact on the thought of his time. His free-verse odes, which so erotically fused the body and the soul, ac-

---

* Sadly, it would take another thirty years — and another brutal war — before sensory ghosts were rediscovered. In 1917, confronted by the maimed soldiers of WWI, the neurologist J. Babinski described his own version of sensory ghosts. He makes no mention of Herman Melville, William James, or Weir Mitchell.

tually precipitated a parallel discovery within psychology. An avid Whitman enthusiast, William James was the first scientist to realize that Whitman's poetry was literally true: the body was the source of feelings. The flesh was not a part of what we felt, it *was* what we felt. As Whitman had prophetically chanted, "Behold, the body includes and is the meaning, the main concern, and includes and is the soul."

His entire life, James loved reading Whitman's poetry out loud, feeling the "passionate and mystical ontological emotion that suffuses his words." In Whitman, James discovered a "contemporary prophet" able to "abolish the usual human distinctions." According to James, Whitman's poetic investigations of the body had discovered "the kind of fiber . . . which is the material woven of all the excitements, joys, and meanings that ever were, or ever shall be, in this world." Whitman realized how we feel.

The convergent beliefs of James and Whitman should not be surprising. After all, they shared a common source: Emerson. When Emerson came to New York City on his lecture tour in 1842, his speech "The Poet" was lauded in the papers by the journalist Walter Whitman, who would take his line about a "meter making argument" literally. While in the city, Emerson also met with Henry James Sr., a dilettante mystic and critic, and was invited into his New York City home. William James, Henry Sr.'s eldest son, had just been born. Legend has it that Emerson blessed William in his cradle and became the infant's godfather.

True or false, the story accurately reflects the intellectual history of America. William James inherited the philosophical tradition of Emerson. Pragmatism, the uniquely American philosophy James invented, was in part a systematization of Emerson's skeptical mysticism. Like Emerson and Whitman, James always enjoyed puncturing the pretensions of nineteenth-century science. He thought that people should stop thinking of scientific theories as mirrors of nature, what he called "the copy version of truth." Instead, they should see its facts as tools, which "help us get into a satisfactory relation with experience." The truth of an idea, James wrote, is the use of an idea, its "cash-value." Thus, according to the pragmatists, a practical poet could be just as truthful as an accurate experiment. All that mattered was the "concrete difference" an idea produced in our actual lives.

But before he became a philosopher, William James was a psychologist. In 1875, he established one of the world's first psychological laboratories at Harvard. Though he was now part of the medical school, James had no intention of practicing "brass instrument psychology," his critical name for the new scientific approach that tried to quantify the mind in terms of its elemental sensations. What physicists had done for the universe, these psychologists wanted to do for consciousness. Even their vocabulary was stolen straight from physics: thought had a "velocity," nerves had "inertia," and the mind was nothing but its "mechanical reflex-actions." James was contemptuous of such a crude form of reductionism. He thought its facts were useless.

James also wasn't very good at this new type of psychology. "It is a sort of work which appeals particularly to patient and exact minds," he wrote in his masterpiece, *The Principles of Psychology*, and James realized that his mind was neither patient nor particularly exact. He loved questions more than answers, the uncertainty of faith more than the conviction of reason. He wanted to call the universe the pluriverse. In his own psychological experiments, James was drawn to the phenomena that this mental reductionism ignored. What parts of the mind *cannot* be measured?

Searching for the immeasurable led James directly to the question of feeling. Our subjective emotions, he said, were the "unscientific half of existence."* Because we only experienced the feeling as a conscious whole — and not as a sum of separate sensations — to

* As a pragmatist, James also believed that feelings — and not some sort of pure Cartesian reason — were the motivation behind most of our beliefs. In "The Will to Believe," James remarked that "these feelings of our duty about either truth or error are in any case only expressions of our passional life . . . Objective evidence and certitude are doubtless very fine ideals to play with, but where on this moonlit and dream-visited planet are they found?" Although James's essay sparked a firestorm of controversy, he was really just taking David Hume's claim that "reason is, and ought to be, the slave of the passions" to its logical conclusion.

break the emotion apart (as science tried to do) was to make it un-real. "The demand for atoms of feeling," James wrote, "seems a sheer vagary, an illegitimate metaphor. Rationally, we see what per-plexities it brings in train; and empirically, no fact suggests it, for the actual content of our minds are always representations of some kind of *ensemble*."

Ensemble is the key word here. As Whitman had written thirty years before, "I will not make poems with reference to parts / But I will make poems with reference to ensemble." When James intro-spected, he realized that Whitman's poetry revealed an essential truth: our feelings emerge from the interactions of the brain *and* the body, not from any single place in either one. This psychological theory, first described in the 1884 article "What Is an emotion?"* is Whitman, pure and simple. Like Whitman, James concluded that if consciousness was severed from the body, "there would be nothing left behind, no 'mind-stuff' out of which the emotion can be con-stituted." As usual, James's experimental evidence consisted of ordi-nary experience. He structured his argument around vivid exam-ples stolen straight from real life, such as encountering a bear in the woods. "What kind of an emotion of fear," he wondered, "would be left [after seeing the bear] if the feeling of quickened heart beats nor of shallow breathing, neither of trembling lips nor of weakened limbs, neither of goose bumps nor of visceral stirrings, were pres-ent?" James's answer was simple: without the body there would be no fear, for an emotion begins as the perception of a bodily change. When it comes to the drama of feelings, our flesh is the stage.

At first glance, this theory of emotion seems like the height of materialism, a reduction of feeling to a physical state. But James was actually making the opposite point. Inspired by Whitman's poetic sense of unity, James believed that our emotions emerged from the constant interaction of the body and the brain. Just as fear cannot

---

* A year after James's article was published, the Danish psychologist Carl Lange published a similar theory about body and emotion, leading scientists to refer to the theory as the James-Lange hypothesis.

be abstracted from its carnal manifestations, it also cannot be separated from the mind, which endows the body's flesh with meaning. As a result, science cannot define feeling without also taking consciousness — what the feeling is *about* — into account. "Let not this view be called materialistic," James warns his reader. "Our emotions must always be inwardly what they are, whatever be the physiological ground of their apparition. If they are deep, pure, spiritual facts they remain no less deep, pure, spiritual, and worthy of regard on this present sensation theory. They carry their own inner measure of worth with them."

The Body Electric

Modern neuroscience is now discovering the anatomy underlying Whitman's poetry. It has taken his poetic hypothesis — the idea that feelings begin in the flesh — and found the exact nerves and brain regions that make it true. Antonio Damasio, a neuroscientist who has done extensive work on the etiology of feeling, calls this process the body loop. In his view, the mind stalks the flesh; from our muscles we steal our moods.

How does the brain generate our metaphysical feelings from the physical body? According to Damasio, after an "emotive stimulus" (such as a bear) is seen, the brain automatically triggers a wave of changes in the "physical viscera," as the body prepares for action. The heart begins to pound, arteries dilate, the intestines contract, and adrenaline pours into the bloodstream. These bodily changes are then detected by the cortex, which connects them to the scary sensation that caused the changes in the first place. The resulting mental image — an emulsion of thought and flesh, body and soul — is what we feel. It is an idea that has passed through the vessel of the body.

Over the course of his distinguished career, Damasio has chronicled the lives of patients whose brains have been injured and who, as a result, are missing this intricate body-brain connection. Although they maintain full sensory awareness, these patients are unable to translate their sensations into emotions. The pounding of the heart never becomes a feeling of fear. Because the mind is divorced from the flesh, the patient lives in a cocoon of numbness — numb even to his or her own tragedy.

Damasio's research has elaborated on the necessity of our carnal emotions. His conclusions are Whitmanesque. "The body contributes more than life support," Damasio writes. "It contributes a *content* that is part and parcel of the workings of the normal mind." In fact, even when the body does not literally change, the mind creates a feeling by *hallucinating* a bodily change. Damasio calls this the as-if body loop, since the brain acts as if the body were experiencing a real physical event. By imagining a specific bodily state — like a fast heartbeat, or a surge of adrenaline — the mind can induce its own emotions.

One of Damasio's most surprising discoveries is that the feelings generated by the body are an essential element of rational thought. Although we typically assume that our emotions interfere with reason, Damasio's emotionless patients proved incapable of making reasonable decisions. After suffering their brain injuries, all began displaying disturbing changes in behavior. Some made terrible investments and ended up bankrupt; others became dishonest and antisocial; most just spent hours deliberating over irrelevant details. According to Damasio, their frustrating lives are vivid proof that rationality requires feeling, and feeling requires the body. (As Nietzsche put it, "There is more reason in your body than in your best wisdom.")

Of course, it's hard to make generalizations about the brain based on a few neurological patients. In order to understand how the body loop functions in the normal mind, Damasio devised an ingenious experiment he called the gambling task. The experiment went as follows: a subject — the player — was given four decks of cards, two black and two red, and $2,000 worth of play money. Each card told the player that he had either won or lost money. The subject was instructed to turn over a card from one of the four decks and to make as much money as possible.

But the cards weren't distributed at random. Damasio rigged the game. Two of the decks were full of high-risk cards. These decks had bigger payouts ($100), but also contained extravagant monetary punishments ($1,250). The other two decks, by comparison, were staid and conservative. Although they had smaller payouts ($50), they rarely punished the player. If the gamblers only drew from these two decks, they would come out way ahead.

At first, the card-selection process was entirely random. The player had no reason to favor any specific deck, and so they sampled from each pile, searching for money-making patterns. On average, people had to turn over about fifty cards before they began to only draw from the profitable decks. It took about eighty cards before the average experimental subject could explain *why* they favored those decks. Logic is slow.

But Damasio wasn't interested in logic. He was interested in the body. He attached electrodes to the subjects' palms and measured the electrical conductance of their skin. (As Whitman noted in "I Sing the Body Electric," the body *is* electric, our nerves singing with minor voltages.)* In general, higher levels of conductance in the skin signal nervousness. What Damasio found was that after drawing only ten cards, the hand of the experimental subject got "nervous" whenever it reached for one of the negative decks. While the brain had yet to completely understand the game (and wouldn't for another forty cards), the subject's hand "knew" what deck to draw from. Furthermore, as the hand grew increasingly electric, the sub-

---

* At the time of Whitman's writing, there was very little evidence that the body pulsated with charged ions. Luigi Galvani's discovery in the 1780s that frogs' legs twitched when shocked remained hotly disputed. In fact, it was not until 1875, twenty years *after* Whitman first sang of electric bodies, that Richard Caton, a Liverpool physician, discovered that Whitman was right, the nervous system actually conveys electric current. Caton demonstrated this improbable fact by probing directly on the exposed brains of animals with a reflecting galvanometer (a newly invented device that was able to sense the low voltages of neurons).

ject started drawing more and more frequently from the advantageous decks. The unconscious feelings generated by the body preceded the conscious decision. The hand led the mind.

Whitman would have loved this experiment. In the same poem where he declares the body electric, he also exclaims about "the curious sympathy one feels when feeling with the hand." Long before Damasio, Whitman understood that "the spirit receives from the body just as much as it gives to the body." This is why he listened so closely to his flesh: it was the place where his poetry began.

But Whitman also knew that his poems were not simply odes to the material body. This was the mistake that his Victorian critics made: by taking his references to orgasms and organs literally, they missed his true poetic epiphany. The moral of Whitman's verse was that the body wasn't merely a body. Just as leaves of grass grow out of the dirt, feelings grow out of the flesh. What Whitman wanted to show was how these two different substances — the grass and the dirt, the body and the mind — were actually inseparable. You couldn't write poems about one without acknowledging the presence of the other. As Whitman declared, "I will make the poems of materials, for I think they are to be the most spiritual poems."

This faith in the holiness of everything, even the low things, ultimately led Whitman to dispute the facts of science. When the materialists of his time announced that the body was nothing but an evolved machine — there was no soul inside — Whitman reacted with characteristic skepticism. He believed that no matter how much we knew about our physical anatomy, the ineffable would always remain. This is why he wrote poetry. "Hurray for positive science," Whitman wrote. "Gentlemen, to you the first honors always! / Your facts are useful, and yet they are not my dwelling, / I but enter them to an area of my dwelling."

What Emerson said of Montaigne is true of Whitman too: if you cut his words, they will bleed, "for they are vascular and alive." Whitman's poetry describes our anatomical reality. In the mirror of his art, we see the stark fact of our own improbability. Feeling from

A photograph of Walt Whitman in 1891, just a few months before he died. The photograph was taken by the painter Thomas Eakins.

flesh? Soul from body? Body from soul? Our existence makes no sense. We live inside a contradiction. Whitman exposes this truth, and then, in the very next sentence, accepts it. His only answer is that there is no answer. "I and this mystery, here we stand," Whitman once said, and that pretty much says it all.

Yet the acceptance of contradiction has its own consequences. As Randall Jarrell wrote in an essay on Whitman, "When you organize one of the contradictory elements out of your work of art, you are getting rid not just of it, but of the contradiction of which it was a part; and it is the contradictions in works of art which make them able to represent us — as logical and methodical generalizations cannot — our world and our selves, which are also full of contradictions." By trusting his experience, no matter how paradoxical it might seem, Whitman discovered our anatomical reality. Despite the constant calls for his censure, he never doubted the wisdom of his art. "Now I see it is true, what I guess'd at," Whitman wrote in "Song of Myself." What he guessed at, of course, is that the soul is made of flesh.

For a self-described poet of the body, Whitman's own body was in dreadful shape. Although he often bragged about "the exquisite realization of his health," by the time Whitman died, in the early spring of 1892, his health had been damaged by years of neglect and disease. The doctors who performed his autopsy — they began cutting as soon as Thomas Eakins finished making Whitman's death mask — were startled at the state of his insides. His left lung had collapsed, and only an eighth of his right lung seemed to be in workable condition. Tuberculosis, which he had gotten while serving as a nurse during the Civil War, had chronically inflamed his stomach, liver, and kidneys. He had pneumonia. His heart was swollen. In fact, the only organ which still seemed to be functional was Whitman's brain. Just two months earlier, he had finished compiling his final edition of *Leaves of Grass*, which became the "Death-Bed" edition. As usual, he had revised his old poems and continued to write new ones.

What could Whitman have been thinking as he felt his flesh — his trusted muse — slowly abandon him? He began this last *Leaves of Grass* with a new epigraph, written in death's shadow:

> Come, said my soul,
> Such verses for my Body let us write, (for we are one).

These two poignant lines, the first lines in the last version of his only book of poetry, represent the distilled essence of Whitman's philosophy. We are the poem, his poem says, that emerges from the unity of the body and the mind. That fragile unity — this brief parenthesis of being — is all we have. Celebrate it.

Chapter 2

George Eliot

The Biology of Freedom

> Seldom, very seldom, does complete truth belong to any
> human disclosure; seldom can it happen that something
> is not a little disguised or a little mistaken.
>
> — Jane Austen, *Emma*

GEORGE ELIOT WAS A WOMAN of many names. Born Mary Anne
Evans in 1819, the same year as Queen Victoria, she was at different
times in her life Mary Ann Evans, Marian Evans, Marian Evans
Lewes, Mary Ann Cross, and, always in her art, George Eliot. Each
of her names represented a distinct period of her life, reflecting her
slightly altered identity. Though she lived in a time when women
enjoyed few freedoms, Eliot refused to limit her transformations.
She had no inheritance, but she was determined to write. After
moving to London in 1850 to become an essayist and translator,
Eliot decided, at the age of thirty-seven, to become a novelist. Later
that year, she finished her first novella, *The Sad Fortunes of the Reverend Amos Barton*. She signed the story with her new name; she
was now George Eliot.

Why did she write? After finishing her masterpiece *Middlemarch*
(1872), Eliot wrote in a letter that her novels were "simply a set of
experiments in life — an endeavor to see what our thought and
emotion may be capable of." Eliot's reference to "experiments" isn't
accidental; nothing she wrote was. The scientific process, with its

careful blend of empiricism and imagination, fact and theory, was the model for her writing process. Henry James once complained that Eliot's books contained too much science and not enough art. But James misunderstood Eliot's method. Her novels are fiction in the service of truth, "examination[s] of the history of man" under the "varying experiments of time." Eliot always demanded answers from her carefully constructed plots.

And while her realist form touched upon an encyclopedia of subjects, her novels are ultimately concerned with the nature of the individual. She wanted "to pierce the obscurity of the minute processes" at the center of human life. A critic of naïve romanticism, Eliot always took the bleak facts of science seriously. If reality is governed by mechanical causes, then is life just a fancy machine? Are we nothing but chemicals and instincts, adrift in an indifferent universe? Is free will just an elaborate illusion?

These are epic questions, and Eliot wrote epic novels. Her Victorian fiction interweaves physics and Darwin with provincial politics and melodramatic love stories. She forced the new empirical knowledge of the nineteenth century to confront the old reality of human experience. For Eliot, this was the novel's purpose: to give us a vision of ourselves "more sure than shifting theory." While scientists were searching for our biological constraints — they assumed we were prisoners of our hereditary inheritance — Eliot's art argued that the mind was "not cut in marble." She believed that the most essential element of human nature was its malleability, the way each of us can "will ourselves to change." No matter how many mechanisms science uncovered, our freedom would remain.

Social Physics

In Eliot's time, that age of flowering rationality, the question of human freedom became the center of scientific debate. Positivism — a new brand of scientific philosophy founded by Auguste Comte — promised a utopia of reason, a world in which scientific principles perfected human existence. Just as the theological world of myths and rituals had given way to the philosophical world, so would philosophy be rendered obsolete by the experiment and the bell curve. At long last, nature would be deciphered.

The lure of positivism's promises was hard to resist. The intelligentsia embraced its theories; statisticians became celebrities; everybody looked for something to measure. For the young Eliot, her mind always open to new ideas, positivism seemed like a creed whose time had come. One Sunday, she abruptly decided to stop attending church. God, she decided, was nothing more than fiction. Her new religion would be rational.

Like all religions, positivism purported to explain *everything*. From the history of the universe to the future of history, there was no question too immense to be solved. But the first question for the positivists, and in many ways the question that would be their undoing, was the paradox of free will. Inspired by Isaac Newton's theory of gravity, which divined the cause of the elliptical motions found in the heavens, the positivists struggled to uncover a parallel order behind the motions of humans.* According to their depressing philosophy, we were nothing but life-size puppets pulled by invisible strings.

The founder of this "science of humanity" was Pierre-Simon Laplace. The most famous mathematician of his time, Laplace also served as Napoleon's minister of the interior.† When Napoleon asked Laplace why there was not a single mention of God in his five-volume treatise on cosmic laws, Laplace replied that he "had no need of that particular hypothesis." Laplace didn't need God because he believed that probability theory, his peculiar invention, would solve every question worth asking, including the ancient mystery of human freedom.

Laplace got the idea for probability theory from his work on the orbits of planets. But he wasn't nearly as interested in celestial me-

---

* Newton himself wasn't so naïve: "I can calculate the motion of heavenly bodies," he wrote, "but not the madness of people."

† Napoleon fired Laplace after only six weeks. He said Laplace had "carried the idea of the infinitely small into administration."

chanics as he was in human observation of those mechanics. Laplace knew that astronomical measurements rarely measured up to
Newton's laws. Instead of being clocklike, the sky described by astronomers was consistently inconsistent. Laplace, trusting the order
of the heavens over the eye of man, believed this irregularity re-
sulted from human error. He knew that two astronomers plotting
the orbit of the same planet at the same time would differ reliably in
their data. The fault was not in the stars, but in ourselves.

Laplace's revelation was that these discrepancies could be defeated. The secret was to quantify the errors. All one had to do was
plot the differences in observation and, using the recently invented
bell curve, find the most *probable* observation. The planetary orbit
could now be tracked. Statistics had conquered subjectivity.

But Laplace didn't limit himself to the trajectory of Jupiter or the
rotation of Venus. In his book *Essai sur les Probabilités,* Laplace attempted to apply the probability theory he had invented for astron-
omy to a wide range of other uncertainties. He wanted to show that
the humanities could be "rationalized," their ignorance resolved by
the dispassionate logic of math. After all, the principles underlying
celestial mechanics were no different than those underlying social
mechanics. Just as an astronomer is able to predict the future movement of a planet, Laplace believed that before long humanity would
be able to reliably predict its own behavior. It was all just a matter
of computing the data. He called this brave new science "social
physics."

Laplace wasn't only a brilliant mathematician; he was also an as-
tute salesman. To demonstrate how his new brand of numerology
would one day solve everything — including the future — Laplace
invented a simple thought experiment. What if there were an imaginary being — he called it a "demon" — that "could know all the
forces by which nature is animated"? According to Laplace, such a
being would be omniscient. Since everything was merely matter,
and matter obeyed a short list of cosmic laws (like gravity and inertia), knowing the laws meant knowing everything about everything. All you had to do was crank the equations and decipher the

results. Man would finally see himself for "the automaton that he is." Free will, like God, would become an illusion, and we would see that our lives are really as predictable as the planetary orbits. As Laplace wrote, "We must . . . imagine the present state of the universe as the effect of its prior state and as the cause of the state that will follow it. Freedom has no place here."

But just as Laplace and his cohorts were grasping on to physics as the paragon of truth (since physics deciphered our ultimate laws), the physicists were discovering that reality was much more complicated than they had ever imagined. In 1852, the British physicist William Thomson elucidated the second law of thermodynamics. The universe, he declared, was destined for chaos. All matter was slowly becoming heat, decaying into a fevered entropy. According to Thomson's laws of thermodynamics, the very error Laplace had tried to erase — the flaw of disorder — was actually our future.

James Clerk Maxwell, a Scottish physicist who discovered electromagnetism, the principles of color photography, and the kinetic theory of gases, elaborated on Thomson's cosmic pessimism. Maxwell realized that Laplace's omniscient demon actually violated the laws of physics. Since disorder was real (it was even *increasing*), science had fundamental limits. After all, pure entropy couldn't be solved. No demon could know everything.

But Maxwell didn't stop there. While Laplace believed that you could easily apply statistical laws to specific problems, Maxwell's work with gases had taught him otherwise. While the temperature of a gas was wholly determined by the velocity of its atoms — the faster they fly, the hotter the gas — Maxwell realized that velocity was nothing but a statistical average. At any given instant, the individual atoms were actually moving at different speeds. In other words, all physical laws are only *approximations*. They cannot be applied with any real precision to particulars. This, of course, directly contradicted Laplace's social physics, which assumed that the laws of science were universal and absolute. Just as a planet's position could be deduced from the formula of its orbit, Laplace be-

An etching of George Eliot in 1865 by Paul Adolphe Rajon,
after the drawing by Sir Frederick William Burton

lieved, our behaviors could be plotted in terms of our own iron-clad forces. But Maxwell knew that every law had its flaw. Scientific theories were functional things, not perfect mirrors to reality. Social physics was founded on a fallacy.

Love and Mystery

George Eliot's belief in positivism began to fade when she suffered a broken heart. Here was a terrible feeling no logic could solve. The cause of her sadness was Herbert Spencer, the Victorian biologist who coined the phrase "survival of the fittest." After Eliot moved to London, where she lived in a flat on the Strand, she grew intimate with Spencer. They shared long walks in the park and a subscription to the opera. She fell in love. He did not. When he began to ignore her — their relationship was provoking the usual Victorian rumors — Eliot wrote Spencer a series of melodramatic yet startlingly honest love letters. She pleaded for his "mercy and love": "I want to know if you can assure me that you will not forsake me, and that you will always be with me as much as you can and share your thoughts and feelings with me. If you become attached to someone else, then I must die, but until then I could gather courage to work and make life valuable, if only I had you near me." Despite Eliot's confessions of vulnerability, the letter proudly concludes with an acknowledgment of her worth: "I suppose no woman before ever wrote such a letter as this — but I am not ashamed of it, for I am conscious that in the light of reason and true refinement I am worthy of your respect and tenderness."

Spencer ignored Eliot's letters. He was steadfast in his rejection. "The lack of physical attraction was fatal," he would later write, blaming Eliot's famous ugliness for his absence of feeling. He could not look past her "heavy jaw, large mouth, and big nose."* Spencer believed his reaction was purely biological, and was thus immutable: "Strongly as my judgment prompted, my instincts would not respond." He would never love Eliot.

Her dream of marriage destroyed, Eliot was forced to confront a

---

* Shortly after breaking Eliot's heart, Spencer wrote two cruel essays on "Personal Beauty" in which he extolled the virtues of prettiness. In his autobiography he had the audacity to brag of his shallowness, and wrote, in a veiled reference to Eliot, that "Physical beauty is a *sine qua non* with me; as was once unhappily proved where the intellectual traits and the emotional traits were of the highest."

But if Spencer was a man unable to see beyond appearances, Henry James believed Eliot was proof that personality could triumph over prettiness. James memorably described his first meeting with Eliot: "To begin with, she is magnificently ugly — deliciously hideous. She has a low forehead, a dull grey eye, a vast pendulous nose, a huge mouth full of uneven teeth and a chin and jaw bone qui n'en finessent [sic] pas . . . Now in this vast ugliness resides a most powerful beauty which, in a very few minutes steals forth and charms the mind, so that you end as I ended, in falling in love with her."

future as a single, anonymous woman. If she was to support herself, she had to write. But her heartbreak was more than a painful emancipation; it also caused her to think about the world in new ways. In *Middlemarch*, Eliot describes an emotional state similar to what she must have been feeling at the time: "She might have compared her experience at that moment to the vague, alarmed consciousness that her life was taking on a new form, that she was undergoing a metamorphosis . . . Her whole world was in a state of convulsive change; the only thing she could say distinctly to herself was, that she must wait and think anew . . . This was the effect of her loss." In the months following Spencer's rejection, Eliot decided that she would "nourish [a] sleek optimism." She refused to stay sad. Before long, Eliot was in love again, this time with George Henry Lewes.

In many important ways, Lewes was Spencer's opposite. Spencer began his career as an ardent positivist, futilely searching for a theory of everything. After positivism faded away, Spencer became a committed social Darwinist, and he enjoyed explaining all of existence — from worms to civilization — in terms of natural selection. Lewes, on the other hand, was an intellectual renowned for his versatility; he wrote essays on poetry and physics, psychology and philosophy. In an age of increasing academic specialization, Lewes remained a Renaissance man. But his luminous mind concealed a desperate unhappiness. Like Eliot, Lewes was also suffering from a broken heart. His wife, Agnes, was pregnant with the child of his best friend.

In each other, Lewes and Eliot found the solution for their melancholy. Lewes would later describe their relationship as deeply, romantically mysterious. "Love," Lewes wrote, "defies all calculation." "We are not 'judicious' in love; we do not select those whom we 'ought to love,' but those whom we cannot help loving." By the end of the year, Lewes and Eliot were traveling together in Germany. He wanted to be a "poet in science." She wanted to be "a scientific poet."

*   *   *

It is too easy to credit love for the metamorphosis of Eliot's world-view. Life's narratives are never so neat. But Lewes did have an un-mistakable effect on Eliot. He was the one who encouraged her to write novels, silencing her insecurities and submitting her first manuscript to a publisher.

Unlike Spencer, Lewes never trusted the enthusiastic science of the nineteenth century. A stubborn skeptic, Lewes first became fa-mous in 1855 with his *Life of Goethe,* a sympathetic biography that interwove Goethe's criticisms of the scientific method with his ro-mantic poetry. In Goethe, Lewes found a figure who resisted the mechanistic theories of positivism, trusting instead in the "concrete phenomena of *experience.*" And while Lewes eagerly admitted that a properly experimental psychology could offer an "objective insight into our thinking organ," he believed that "Art and Literature" were no less truthful, for they described the "psychological world." In an age of ambitious experiments, Lewes remained a pluralist.

Lewes's final view of psychology, depicted most lucidly in *The Problems of Life and Mind* (a text that Eliot finished after Lewes's death), insisted that the brain would always be a mystery, "for too complex is its unity." Positivists may proselytize their bleak vision, Lewes wrote, but "no thinking man will imagine anything is *ex-plained* by this. Life and Being remain as inaccessible as ever." If nothing else, freedom is a necessary result of our ignorance.

By the time Eliot wrote her last novel, *Daniel Deronda* (1876), she had come to see that Laplace and Spencer and the rest of the positivists were wrong. The universe could not be distilled into a neat set of causes. Freedom, however fragile, exists. "Necessitarian-ism,"* Eliot wrote, "I hate the ugly word." Eliot had read Maxwell on molecules, even copying his lectures into her journals, and she knew that nothing in life could be perfectly predicted. To make her point, Eliot began *Daniel Deronda* with a depiction of human be-

---

* Necessitarianism is a synonym for determinism. Popular in the nineteenth century, the theory holds that human actions are "necessitated" by antecedent causes over which we have no control.

ings as imagined by Laplace. The setting is a hazy and dark casino, full of sullen people who act, Eliot writes, "as if they had all eaten of some root that for the time compelled the brains of each to the same narrow monotony of action." These gamblers are totally powerless, dependent on the dealer to mete out their random hands. They passively accept whichever cards they are dealt. Their fortune is determined by the callous laws of statistics.

In Eliot's elaborately plotted work, the casino is no casual prop — it is a criticism of determinism. As soon as Eliot introduces this mechanical view of life, she begins deconstructing its silly simplicities. After Daniel enters the casino, he spies a lone woman, Gwendolen Harleth. "Like dice in mid-air," Gwendolen is an unknown. Her mysteriousness immediately steals Daniel's attention; she transcends the depressing atmosphere of the casino. Unlike the gamblers, who do nothing but wait for chance to shape their fate, Gwendolen seems free. Daniel stares at her and wonders: "Was she beautiful or not beautiful? And what was the secret of form or expression which gave the dynamic quality to her glance?"

Eliot uses the casino to remind us that we are also mysterious, a "secret of form." And because Gwendolen is a dynamic person, her own "determinate," she will decide how her own life unfolds. Even when she is later entrapped in a marriage to the evil Grandcourt — "his voice the power of thumbscrews and the cold touch of the rack" — she manages to free herself. Eliot creates Gwendolen to remind us that human freedom is innate, for we are the equation without a set answer. We solve ourselves.*

While George Eliot spurned the social physics of her day, she greeted Darwin's theory of natural selection as the start of a new "epoch." She read *On the Origin of Species* when it was first published in 1859

---

* Of course, being free also makes us accountable for our behavior. One of Eliot's main problems with social physics was that it denied humanity any moral agency. After all, if every action has an external cause, then it seems cruel to punish cruelty. In her novels, Eliot wanted to describe a more realistic vision of human nature and thus inspire us to become better. If social physics made us callous, then art might make us compassionate.

and immediately realized that the history of life now had a coherent structure. Here was an authentic version of our beginning. And while positivists believed that the chaos of life was only a façade, that beneath everything lay the foundation of physical order, Darwinism said that randomness was a fact of nature. In many ways, randomness was *the* fact of nature.* According to Darwin, in a given population sheer chance dictated variety. Genetic mutations (Darwin called them *saltations*) followed no natural laws. This diversity created differing rates of reproduction among organisms, which led to the survival of the fittest. Life progressed *because* of disorder, not despite it. The theologian's problem — the question of why nature contained so much suffering and contingency — became Darwin's solution.

The bracing embrace of chance was what attracted Eliot to Darwin. Here was a narrative that was itself unknowable, since it was guided by random variation. The evolution of life depended on events that had no discernible cause. Unlike Herbert Spencer, who believed that Darwin's theory of evolution could solve *every* biological mystery (natural selection was the new social physics), Eliot believed that Darwin had only deepened the mystery. As she confided to her diary: "So the world gets on step by step towards brave clearness and honesty! But to me the Development theory [Darwin's theory of evolution] and all other explanations of processes by which things came to be produce a feeble impression compared with the mystery that lies under the process." Because evolution has no purpose or plan — it is merely the sum of its accumulated mistakes — our biology remains impenetrable. "Even Science, the strict measurer," Eliot confessed, "is obliged to start with a make-believe unit."

The intrinsic mystery of life is one of Eliot's most eloquent themes. Her art protested against the braggadocio of positivism, which assumed that everything would one day be defined by a few omnipo-

---

* Darwin acknowledged the deep chanciness at the heart of natural selection. Although he never uses the adjective *random*, Darwin constantly asserts that variations are "undirected" and "occur in no determinate way."

tent equations. Eliot, however, was always most interested in what we couldn't know, in those aspects of reality that are ultimately irreducible: "If we had a keen vision and feeling of *all* ordinary human life," she warns us in *Middlemarch*, "it would be like hearing the grass grow and the squirrel's heart beat, and we should die of that roar which lies on the other side of silence. As it is, the quickest of us walk about well wadded with stupidity." Those characters in her novels who deny our mystery, who insist that freedom is an illusion and that reality is dictated by abstract laws (which they happen to have discovered), work against the progress of society. They are the villains, trusting in "inadequate ideas." Eliot was fond of quoting Tennyson's *In Memoriam:* "There lives more faith in honest doubt, / Believe me, than in half the creeds."

Middlemarch, Eliot's masterpiece, contains two reductionists searching for what Laplace called "the final laws of the world." Edward Casaubon, the pretentious husband of Dorothea Brooke, spends his days writing a "Key to All Mythologies," which promises to find the hidden connection between the varieties of religious experience. His work is bound to fail, Eliot writes, for he is "lost among small closets and winding stairs." Casaubon ends up dying of a "fatty degeneration of the heart," a symbolic death if ever there was one.

Dr. Tertius Lydgate, the ambitious country doctor, is engaged in an equally futile search, looking for the "primitive tissue of life." His foolish quest is an allusion to Herbert Spencer's biological theories, which Eliot enjoyed mocking.* Like Casaubon, Lydgate continually overestimates the explanatory power of his science. But reality eventually intrudes and Lydgate's scientific career collapses. After enduring a few financial mishaps, Lydgate ends up becoming a doctor of gout, and "considers himself a failure: he had not done

---

* Eliot enjoyed telling a story about her botanical expeditions to Kew Gardens with Spencer. Being a devout Darwinist, Spencer explained the structure of every flower by referencing some vague story about "the necessity of evolutionary development." But if a flower failed to fit his neat theory, then it was *"tant pis pour les fleurs"* ("too bad for the flowers").

what he once meant to do." His own life becomes a testament to the limits of science.

After Casaubon dies, Dorothea, the heroine of *Middlemarch*, who bears an uncanny resemblance to Eliot, falls in love with Will Ladislaw, a poetic type and not-so-subtle symbol of free will. (Will is in "passionate rebellion against his inherited blot.") Tragically, because of Casaubon's final will (notice the emerging theme), Dorothea is unable to act on her love. If she marries Will, who is of low social rank, she loses her estate. And so she resigns herself to a widowed unhappiness. Many depressing pages ensue. But then Will returns to Middlemarch, and Dorothea, awakened by his presence, realizes that she wants to be with him. Without freedom, money is merely paper. She renounces Casaubon's estate and runs away with her true love. Embracing Will is her first act of free will. They live happily ever after, in "the realm of light and speech."

But *Middlemarch*, a novel that denies all easy answers, is more complicated than its happy ending suggests. (Virginia Woolf called *Middlemarch* "one of the few English novels written for grown-up people.") Eliot had read too much Darwin to trust in the lasting presence of joy. She admits that each of us is born into a "hard, unaccommodating Actual." This is why Dorothea, much to Eliot's dismay, could not end the novel as a single woman. She was still trapped by the social conventions of the nineteenth century. As Eliot admonishes in the novel's final paragraphs, "There is no creature whose inward being is so strong that it is not greatly determined by what lies outside it."

In her intricate plots, Eliot wanted to demonstrate how the outside and the inside, our will and our fate, are in fact inextricably entangled. "Every limit is a beginning as well as an ending," Eliot confesses in *Middlemarch*. Our situation provides the raw material out of which we make our way, and while it is important "never to beat and bruise one's wings against the inevitable," it is always possible "to throw the whole force of one's soul towards the achievement of some possible better." You can always change your life.

The Brand-New Mind

If science could *see* freedom, what would it look like? If it wanted to find the will, where would it search? Eliot believed that the mind's ability to alter itself was the source of our freedom. In *Middlemarch*, Dorothea — a character who, like Eliot herself, never stopped changing — is reassured that the mind "is not cut in marble — it is not something solid and unalterable. It is something living and changing." Dorothea finds hope in this idea, since it means that the soul "may be rescued and healed." Like Jane Austen, a literary forebear, Eliot reserved her highest praise for characters brave enough to embrace the possibilities of change. Just as Elizabeth Bennet escapes her own prejudices, so does Dorothea recover from her early mistakes. As Eliot wrote, "we are a process and an unfolding."

Biology, at least until very recently, did not share Eliot's faith in the brain's plasticity. While Laplace and the positivists saw our environment as a prison — from its confines, there was no escape — in the time after Darwin, determinism discovered a new stalking-horse. According to biology, the brain was little more than a genetically governed robot, our neural connections dictated by forces beyond our control. As Thomas Huxley disdainfully declared, "We are conscious automata."

The most glaring expression of that theme was the scientific belief that a human was born with a complete set of neurons. This theory held that brain cells — unlike every other cell in our body — didn't divide. Once infancy was over, the brain was complete; the fate of the mind was sealed. Over the course of the twentieth century, this idea became one of neuroscience's fundamental principles.

The most convincing defender of this theory was Pasko Rakic, of Yale University. In the early 1980s, Rakic realized that the idea that neurons never divide had never been properly tested in primates. The dogma was entirely theoretical. Rakic set out to investigate. He studied twelve rhesus monkeys, injecting them with radioactive thymidine, which allowed him to trace the development of neurons in the brain. Rakic then killed the monkeys at various stages after the injection of the thymidine and searched for signs of new neurons. There were none. "All neurons of the rhesus monkey brain are generated during pre-natal and early post-natal life," Rakic wrote in his influential paper "Limits of Neurogenesis in Primates," which he published in 1985. While Rakic admitted that his proof wasn't perfect, he persuasively defended the dogma. He even went so far as to construct a plausible evolutionary theory as to why neurons couldn't divide. Rakic imagined that at some point in our distant past, primates had traded the ability to give birth to new neurons for the ability to modify the connections between our old neurons. According to Rakic, the "social and cognitive" behavior of primates *required* the absence of neurogenesis. His paper, with its thorough demonstration of what everyone already believed, seemed like the final word on the matter. His experiments were never independently verified.

The genius of the scientific method, however, is that it accepts no permanent solution. Skepticism is its solvent, for every theory is imperfect. Scientific facts are meaningful precisely because they are ephemeral, because a new observation, a more honest observation, can always alter them. This is what happened to Rakic's theory of the fixed brain. It was, to use Karl Popper's verb, *falsified*.

In 1989, Elizabeth Gould, a young postdoc working in the lab of Bruce McEwen at Rockefeller University, in New York City, was investigating the effect of stress hormones on rat brains. Chronic stress is devastating to neurons, and Gould's research focused on the death of cells in the hippocampus. But while Gould was documenting the brain's degeneration, she happened upon something completely miraculous: the brain also healed itself.

Confused by this anomaly, Gould went to the library. She assumed she was making some simple experimental mistake, because *neurons don't divide.* Everybody knew that. But then, looking through a dusty twenty-seven-year-old science journal, Gould found a tantalizing clue. Beginning in 1962, a researcher at MIT, Joseph Altman, published several papers claiming that adult rats, cats, and guinea pigs all formed new neurons. Although Altman used the same technique that Rakic later used in monkey brains — the injection of radioactive thymidine — his results were ridiculed, and then ignored.

As a result, the brand-new field of neurogenesis vanished before it began. It would take another decade before Michael Kaplan, at the University of New Mexico, would use an electron microscope to image neurons giving birth to new neurons. Kaplan discovered these fresh cells everywhere in the mammalian brain, including the cortex. Yet even with this visual evidence, science remained stubbornly devoted to its doctrine. After enduring years of scorn and skepticism, Kaplan, like Altman before him, abandoned the field of neurogenesis.

Reading Altman's and Kaplan's papers, Gould realized that her mistake wasn't a mistake: it was an ignored fact. The anomaly had been suppressed. But the final piece of the puzzle came when Gould discovered the work of Fernando Nottebohm, who was, coincidentally, also at Rockefeller. Nottebohm, in a series of remarkably beautiful studies on bird brains, showed that neurogenesis was required for bird song. To sing their complex melodies, male birds needed new brain cells. In fact, up to 1 percent of the neurons in the bird's song center were made fresh every day. "At the time, this was a radical idea," Nottebohm says. "The brain was thought to be a very fixed organ. Once development was over, scientists assumed that the mind was cast in a crystalline structure. That was it; you were done."

Nottebohm disproved this dogma by studying birds in their actual habitat. If he had kept his birds in metal cages, depriving them of their natural social context, he would never have observed the abundance of new cells that he did. The birds would have been too stressed to sing, and fewer new neurons would have been created. As Nottebohm has said, "Take nature away and all your insight is in a biological vacuum." It was only because he looked at birds *outside* of the laboratory's vacuum that he was able to show that neurogenesis, at least in finches and canaries, had a real evolutionary purpose.

Despite the elegance of Nottebohm's data, his science was marginalized. Bird brains were seen as irrelevant to the mammalian brain. Avian neurogenesis was explained away as an exotic adaptation, a reflection of the fact that flight required a light cerebrum. In his *Structure of Scientific Revolutions,* the philosopher of science Thomas Kuhn wrote about how science tends to exclude its contradictions: "Until the scientist has learned to see nature in a different way, the new fact is not quite a scientific fact at all." Evidence of neurogenesis was systematically excluded from the world of "normal science."

But Gould, motivated by the strangeness of her own experimental observations, connected the dots. She realized that Altman, Kaplan, and Nottebohm all had strong evidence for mammalian neurogenesis. Faced with this mass of ignored data, Gould abandoned her earlier project and began investigating the birth of neurons.

She spent the next eight years quantifying endless numbers of radioactive rat brains. But the tedious manual labor paid off. Gould's data shifted the paradigm. More than thirty years had passed since Altman first glimpsed new neurons, but neurogenesis had finally become a scientific fact.

After her frustrating postdoc, during which time her science was continually attacked, Gould was offered a job at Princeton. The very next year, in a series of landmark papers, she began documenting neurogenesis in primates, in direct contradiction of Rakic's data. She demonstrated that marmosets and macaques created new neurons throughout life. The brain, far from being fixed, is actually in a constant state of cellular upheaval. By 1998, even Rakic admitted that neurogenesis was real, and he reported seeing new neurons in rhesus monkeys.* The textbooks were rewritten: the brain is constantly giving birth to itself.

Gould has gone on to show that the amount of neurogenesis is itself modulated by the environment, and not just by our genes. High levels of stress can decrease the number of new cells; so can being low in a dominance hierarchy (the primate equivalent of being low class). In fact, monkey mothers who live in stressful conditions give birth to babies with drastically reduced neurogenesis, even if those babies never experienced stress themselves. But there is hope: the scars of stress can be healed. When primates were transferred to enriched enclosures — complete with branches, hidden food, and a rotation of toys — their adult brains began to recover rapidly. In less than four weeks, their deprived cells underwent radical renovations and formed a wealth of new connections. Their rates of neurogenesis returned to normal levels. What does this data mean? The mind is never beyond redemption, for no environment can extinguish neurogenesis. As long as we are alive, important parts of the brain are dividing. The brain is not marble, it is clay, and our clay never hardens.

Neuroscience is just beginning to explore the profound ramifications of this discovery. The hippocampus, the part of the brain that modulates learning and memory, is continually supplied with new neurons, which help us to learn and remember new ideas and behaviors. Other scientists have discovered that antidepressants work by stimulating neurogenesis (at least in rodents), implying that depression is ultimately caused by a decrease in the amount

---

* How did Rakic make his original mistake? There is no easy answer. Rakic is an excellent scientist, one of the finest neuroscientists of his generation. But seeing radioactive new neurons is extremely difficult. These cells are easy to ignore, especially if they *shouldn't* be there. One has to be looking for them in order to see them. Furthermore, almost all lab primates live in an environment that suppresses neurogenesis. A drab-looking cage creates a drab-looking brain. Unless the primates are transferred to an enriched enclosure, their adult brains will produce few new neurons. The realization that typical laboratory conditions are debilitating for animals and produce false data has been one of the accidental discoveries of the neurogenesis field.

of new neurons, and not by a lack of serotonin. A new class of anti-depressants is being developed that targets the neurogenesis pathway. For some reason, newborn brain cells make us happy.

And while freedom remains an abstract idea, neurogenesis is cellular evidence that we evolved to never stop evolving. Eliot was right: to be alive is to be ceaselessly beginning. As she wrote in *Middlemarch*, the "mind [is] as active as phosphorus." Since we each start every day with a slightly new brain, neurogenesis ensures that we are never done with our changes. In the constant turmoil of our cells — in the irrepressible plasticity of our brains — we find our freedom.

The Literary Genome

Even as neuroscience began to reveal the brain's surprisingly supple structure, other scientists were becoming entranced with an even more powerful deterministic principle: genetics. When James Watson and Francis Crick discovered the chemical structure of DNA, in 1953, they gave biology a molecule that seemed to explain life itself. Here was our source stripped bare, the incarnate reduced to some nucleic acids and weak hydrogen bonds. Watson and Crick recognized the handsome molecule the moment they assembled it out of their plastic atoms. What they had constructed was a double helix, a spiraling structure composed of two interwoven threads. The form of the double helix suggested how it might convey its genetic information. The same base pairs that held the helix together also represented its code, a hieroglyph consisting of four letters: A, T, C, and G.

Following Watson and Crick, scientists discovered how the primitive language of DNA spelled out the instructions for complex organisms. They summarized the idea in a simple epithet, the Central Dogma: DNA made RNA that made protein. Since we were merely elaborate sculptures of protein, biologists assumed that we were the sum of our DNA. Crick formulated the idea this way: "Once 'information' has passed into the protein [from the DNA,] it *can-*

not get out again." From the perspective of genetics, life became a neat causal chain, our organism ultimately reducible to its text, these wispy double helices afloat in the cellular nuclei. As Richard Dawkins declared in *The Selfish Gene,* "We are survival machines — robot vehicles blindly programmed to preserve the selfish molecules known as genes."

The logical extension of this biological ideology was the Human Genome Project. Begun in 1990, the project was an attempt to decode the genetic narrative of our species. Every chromosome, gene, and base pair would be sequenced and understood. Our textual underpinnings would be stripped of their mystery, and our lack of freedom would finally be exposed. For the paltry sum of $2.7 billion, everything from cancer to schizophrenia would be eradicated.

That was the optimistic hypothesis. Nature, however, writes astonishingly complicated prose. If our DNA has a literary equivalent, it's *Finnegans Wake.* As soon as the Human Genome Project began decoding our substrate, it was forced to question cherished assumptions of molecular biology. The first startling fact the project uncovered was the dizzying size of our genome. While we technically need only 90 million base pairs of DNA to encode the 100,000 different proteins in the human body, we actually have more than 3 billion base pairs. Most of this excess text is junk. In fact, more than 95 percent of human DNA is made up of what scientists call introns: vast tracts of repetitive, noncoding nonsense.

But by the time the Human Genome Project completed its epic decoding, the dividing line between genes and genetic filler had begun to blur. Biology could no longer even define what a gene was. The lovely simplicity of the Central Dogma collapsed under the complications of our genetic reality, in which genes are spliced, edited, methylated, and sometimes jump chromosomes (these are called *epi*genetic effects). Science had discovered that, like any work of literature, the human genome is a text in need of commentary, for what Eliot said of poetry is also true of DNA: "all meanings depend on the key of interpretation."

What makes us human, and what makes each of us his or her

own human, is not simply the genes that we have buried in our base pairs, but how our cells, in dialogue with our environment, feed back to our DNA, changing the way we read ourselves. Life is a dialectic. For example, the code sequence GTAAGT can be translated as instructions to insert the amino acid valine and serine; read as a spacer, a genetic pause that keeps other protein parts an appropriate distance from one another; or interpreted as a signal to cut the transcript at that point. Our human DNA is defined by its multiplicity of possible meanings; it is a code that requires context. This is why we can share 42 percent of our genome with an insect and 98.7 percent with a chimpanzee and yet still be so completely different from both.

By demonstrating the limits of genetic determinism, the Human Genome Project ended up becoming an ironic affirmation of our individuality. By failing to explain us, the project showed that humanity is not simply a text. It forced molecular biology to focus on how our genes interact with the real world. Our nature, it turns out, is endlessly modified by our nurture. This uncharted area is where the questions get interesting (and inextricably difficult).

Take the human mind. If its fissured cortex — an object that is generally regarded as the most complicated creation in the known universe — were genetically programmed, then it should have many more genes than, say, the mouse brain. But this isn't the case. In fact, the mouse brain contains roughly the same number of genes as the human brain. After decoding the genomes of numerous species, scientists have found that there is little correlation between genome size and brain complexity. (Several species of amoeba have much larger genomes than humans.) This strongly suggests that the human brain does not develop in accordance with a strict genetic program that specifies its design.

But if DNA doesn't determine the human brain, then what does? The easy answer is: nothing. Although genes are responsible for the gross anatomy of the brain, our plastic neurons are designed to adapt to our experiences. Like the immune system, which alters itself in response to the pathogens it actually encounters (we do not have the B cells of our parents), the brain is constantly adapting to the particular conditions of life. This is why blind people can use their visual cortex to read Braille, and why the deaf can process sign language in their auditory cortex. Lose a finger and, thanks to neural plasticity, your other fingers will take over its brain space. In one particularly audacious experiment, the neuroscientist Mriganka Sur literally rewired the mind of a ferret, so that the information from its retina was plugged into its auditory cortex. To Sur's astonishment, the ferrets could still see. Furthermore, their auditory cortex now resembled the typical ferret visual cortex, complete with spatial maps and neurons tuned to detect slants of light. Michael Merzenich, one of the founders of the plasticity field, called this experiment "the most compelling demonstration you could have that experience shapes the brain." As Eliot always maintained, the mind is defined by its malleability.*

This is the triumph of our DNA: it makes us without determining us. The invention of neural plasticity, which is encoded by the genome, lets each of us transcend our genome. We *emerge*, characterlike, from the vague alphabet of our text. Of course, to accept the freedom inherent in the human brain — to know that the individual is *not* genetically predestined — is also to accept the fact that we have no single solutions. Every day each one of us is given the gift of new neurons and plastic cortical cells; only we can decide what our brains will become.

The best metaphor for our DNA is literature. Like all classic liter-

---

* Scientists are now discovering that even mental traits that have a strong genetic component — such as IQ — are incredibly sensitive to changes in the environment. A French study of troubled children adopted between the ages of four and six clearly demonstrated the way our innate human nature depends on how we are nurtured. At the time they were adopted, these young children had IQs that averaged around 77, putting them near retardation level. However, when the children retook the IQ test nine years later, all of them did significantly better. This was extremely surprising, since IQ is supposed to be essentially stable over the course of a lifetime. Furthermore, the amount that a child improved was directly related to the adopting family's socioeconomic status. Children adopted by middle-class families had average scores of 92; those placed in upper-class homes had their IQs climb, on average, more than 20 points, to 98. In a relatively short time, their IQs had gone from significantly below average to practically normal.

ary texts, our genome is defined not by the certainty of its meaning, but by its linguistic instability, its ability to encourage a multiplicity of interpretations. What makes a novel or poem immortal is its innate complexity, the way every reader discovers in the same words a different story. For example, many readers find the ending of *Middlemarch,* in which Dorothea elopes with Will, to be a traditional happy ending, in which marriage triumphs over evil. However, some readers — such as Virginia Woolf — see Dorothea's inability to live alone as a turn of plot "more melancholy than tragedy." The same book manages to inspire two completely different conclusions. But there is no right interpretation. Everyone is free to find his or her own meaning in the novel. Our genome works the same way. Life imitates art.

The Blessing of Chaos

How does our DNA inspire such indeterminacy? After all, *Middlemarch* had an author; she deliberately crafted an ambiguous ending. But real life doesn't have an intelligent designer. In order to create the wiggle room necessary for individual freedom, natural selection came up with an ingenious, if unnerving, solution. Although we like to imagine life as a perfectly engineered creation (our cells like little Swiss clocks), the truth is that our parts aren't predictable. Bob Dylan once said, "I accept chaos. I'm not sure whether it accepts me." Molecular biology, confronted with the unruliness of life, is also forced to accept chaos. Just as physics discovered the indeterminate quantum world — a discovery that erased classical notions about the fixed reality of time and space — so biology is uncovering the unknowable mess at its core. Life is built on an edifice of randomness.

One of the first insights into the natural disorder of life arrived in 1968, when Motoo Kimura, the great Japanese geneticist, introduced evolutionary biology to his "neutral theory of molecular evolution." This is a staid name for what many scientists consider the most interesting revision of evolutionary theory since Darwin.

Kimura's discovery began with a paradox. Starting in the early 1960s, biologists could finally measure the rate of genetic change in species undergoing natural selection. As expected, the engine of evolution was random mutation: double helices suffered from a constant barrage of editing errors. Buried in this data, however, was an uncomfortable new fact: DNA changes way too much. According to Kimura's calculations, the average genome was changing at a hundred times the rate predicted by the equations of evolution. In fact, DNA was changing so much that there was no possible way natural selection could account for all of these so-called adaptations.

But if natural selection wasn't driving the evolution of our genes, then what was? Kimura's answer was simple: chaos. Pure chance. The dice of mutation and the poker of genetic drift. At the level of our DNA, evolution works mostly by accident.* Your genome is a record of random mistakes.

But perhaps that randomness is confined to our DNA. The clocklike cell must restore some sense of order, right? Certainly the *translation* of our genome — the expression of our actual genes — is a perfectly regulated process, with no hint of disarray. How else could we function? Although molecular biology used to assume that was the case, it isn't. Life is slipshod. Inside our cells, shards and scraps of nucleic acid and protein float around aimlessly, waiting to interact. There is no guiding hand, no guarantee of exactness.

In a 2002 *Science* paper entitled "Stochastic Gene Expression in a Single Cell," Michael Elowitz of Caltech demonstrated that biological "noise" (a scientific synonym for chaos) is inherent in gene

---

* Although Kimura's conclusions provoked a storm of controversy — some neo-Darwinists said he was just a creationist with fancy mathematics — they shouldn't have. In fact, Darwin probably would have agreed with Kimura. In the last edition of *On the Origin of Species*, published in 1872, Darwin makes his own position crystal clear: "As my conclusions have lately been much misrepresented, and it has been stated that I attribute the modification of species exclusively to natural selection, I may be permitted to remark that in the first edition of this work, and subsequently, I placed in a most conspicuous position . . . the following words: 'I am convinced that natural selection has been the main, but *not the exclusive* means of modification.' This has been of no avail. Great is the power of steady misinterpretation" [Darwin, 1872, p. 395].

expression. Elowitz began by inserting two separate sequences of DNA stolen from fireflies into the genome of *E. coli.* One gene encoded a protein that made the creatures glow neon green. The other gene made the bacteria radiate red. Elowitz knew that if the two genes were expressed equally in the *E. coli* (as classical biological theory predicted), the color yellow would dominate (for light waves, red plus green equals yellow). That is, if life were devoid of intrinsic noise, all the bacteria would be colored by the same neon hue.

But Elowitz discovered that when the red- and green-light genes were expressed at ordinary levels, and not overexpressed, the noise in the system suddenly became visible. Some bacteria were yellow (the orderly ones), but other cells, influenced by their intrinsic disorder, glowed a deep turquoise or orange. All the variance in color was caused by an inexplicable variance in fluorescent-protein level: the two genes were *not* expressed equally. The simple premise underlying every molecular biology experiment — that life follows regular rules, that it transcribes its DNA faithfully and accurately — vanished in the colorful collage of prokaryotes. Although the cells were technically the same, the randomness built into their system produced a significant amount of fluorescent variation. This disparity in bacterial hue was not reducible. The noise had no single source. It was simply *there*, an essential part of what makes life living.

Furthermore, this messiness inherent in gene translation percolates upward, infecting and influencing all aspects of life. Fruit flies, for example, have long hairs on their bodies that serve as sensory organs. The location and density of those hairs differ between the two sides of the fly, but not in any systematic way. After all, the two sides of the fly are encoded by the same genes and have developed in the same environment. The variation in the fly is a consequence of random atomic jostling inside its cells, what biologists call "developmental noise." (This is also why your left hand and right hand have different fingerprints.)

This same principle is even at work in our brain. Neuroscientist

Fred Gage has found that retrotransposons — junk genes that randomly jump around the human genome — are present at unusually high numbers in neurons. In fact, these troublemaking scraps of DNA insert themselves into almost 80 percent of our brain cells, arbitrarily altering their genetic program. At first, Gage was befuddled by this data. The brain seemed intentionally destructive, bent on dismantling its own precise instructions. But then Gage had an epiphany. He realized that all these genetic interruptions created a population of perfectly unique minds, since each brain reacted to retrotransposons in its own way. In other words, chaos creates individuality. Gage's new hypothesis is that all this mental anarchy is adaptive, as it allows our genes to generate minds of almost infinite diversity.

And diversity is a good thing, at least from the perspective of natural selection. As Darwin observed in *On the Origin of Species,* "The more diversified the descendants from any one species become in structure, constitution and habits, by so much will they be better enabled to seize on many and widely diversified places in the polity of nature." Our psychology bears out this evolutionary logic. From the moment of conception onward, our nervous system is designed to be an unprecedented invention. Even identical twins with identical DNA have strikingly dissimilar brains. When sets of twins perform the same task in a functional MRI machine, different parts of each cortex become activated. If adult twin brains are dissected, the details of their cerebral cells are entirely unique. As Eliot wrote in the preface to *Middlemarch,* "the indefiniteness remains, and the limits of variation are really much wider than anyone would imagine."

Like the discovery of neurogenesis and neural plasticity, the discovery that biology thrives on disorder is paradigm-shifting. The more science knows about life's intricacies, about how DNA actually builds proteins and about how proteins actually build us, the less life resembles a Rolex. Chaos is everywhere. As Karl Popper once said, life is not a clock, *it is a cloud*. Like a cloud, life is "highly irregular, disorderly, and more or less unpredictable." Clouds, crafted and carried by an infinity of currents, have inscrutable wills; they seethe and tumble in the air and are a little different with every moment in time. We are the same way. As has happened so many times before in the history of science, the idée fixe of deterministic order proved to be a mirage. We remain as mysteriously free as ever.

The lovely failure of every reductionist attempt at "solving life" has proved that George Eliot was right. As she famously wrote in 1856, "*Art is the nearest thing to life;* it is a mode of amplifying experience." The sprawling realism of Eliot's novels ended up discovering our reality. We are imprisoned by no genetic or social physics, for life is not at all like a machine. Each of us is free, for the most part, to live as we choose to, blessed and burdened by our own elastic nature. Although this means that human nature has no immutable laws, it also means that we can always improve ourselves, for we are works in progress. What we need now is a new view of life, one that reflects our indeterminacy. We are neither fully free nor fully determined. The world is full of constraints, but we are able to make our own way.

This is the complicated existence that Eliot believed in. Although her novels detail the impersonal forces that influence life, they are ultimately celebrations of self-determination. Eliot criticized all scientific theories that disrespected our freedom, and instead believed "that the relations of men to their neighbours may be settled by algebraic equations." "But," she wrote, "none of these diverging mistakes can co-exist with a real knowledge of the people." What makes humans unique is that each of us is unique. This is why Eliot always argued that trying to define human nature was a useless endeavor, dangerously doomed to self-justification. "I refuse," she wrote, "to adopt any formula which does not get itself clothed for me in some human figure and individual experience." She knew that we inherit minds that let us escape our inheritance; we can always impose our will onto our biology. "I shall not be satisfied with your philosophy," she wrote to a friend in 1875, "till you have conciliated Necessi-tarianism . . . with the practice of willing, of willing to will strongly, and so on."

As Eliot anticipated, our freedom is built into us. At its most fundamental level, life is full of leeway, defined by a plasticity that defies every determinism. We are only chains of carbon, but we transcend our source. Evolution has given us the gift of infinite individuality. There is grandeur in this view of life.

Chapter 3

Auguste Escoffier

The Essence of Taste

> So it happens that when I write of hunger, I am really
> writing about love and the hunger for it, and warmth and
> the love of it and the hunger for it . . . and then the warmth
> and richness and fine reality of hunger satisfied . . . and
> it is all one.
>
> — M.F.K. Fisher, *The Gastronomical Me*

AUGUSTE ESCOFFIER INVENTED veal stock. Others had boiled
bones before, but no one had codified the recipe. Before Escoffier,
the best way to make veal stock was cloaked in mystery; cooking
was like alchemy, a semimystical endeavor. But Escoffier came of
age during the late stages of positivism, a time when knowledge —
some of it true, some of it false — was disseminated at a dizzying
rate. Encyclopedias were the books of the day. Escoffier took this
scientific ethos to heart; he wanted to do for fancy food what
Lavoisier had done for chemistry, and replace the old superstitions
of the kitchen with a new science of cooking.

At the heart of Escoffier's insight (and the source of more than
a few heart failures) was his use of stock. He put it in everything.
He reduced it to gelatinous jelly, made it the base of pureed soups,
and enriched it with butter and booze for sauces. While French
women had created homemade stock for centuries — pot-au-feu
(beef boiled in water) was practically the national dish — Escoffier

gave their protein broth a professional flair. In the first chapter of his *Guide Culinaire* (1903), Escoffier lectured cooks on the importance of extracting flavor from bones: "Indeed, stock is everything in cooking. Without it, nothing can be done. If one's stock is good, what remains of the work is easy; if, on the other hand, it is bad or merely mediocre, it is quite hopeless to expect anything approaching a satisfactory meal." What every other chef was throwing away — the scraps of tendon and oxtail, the tops of celery, the ends of onion, and the irregular corners of carrot — Escoffier was simmering into sublimity.

Although Escoffier introduced his *Guide Culinaire* with the lofty claim that his recipes were "based upon the modern science of gastronomy," in reality he ignored modern science. At the time, scientists were trying to create a prissy nouvelle cuisine based on their odd, and totally incorrect, notions of what was healthy. Pig blood was good for you. So was tripe. Broccoli, on the other hand, caused indigestion. The same with peaches and garlic. Escoffier ignored this bad science (he invented peach Melba), and sautéed away to his heart's malcontent, trusting the pleasures of his tongue over the abstractions of theory. He believed that the greatest threat to public health was the modern transformation of dining from a "pleasurable occasion into an unnecessary chore."

The form of Escoffier's encyclopedic cookbook reflects his romantic bent. Although he was fond of calling *sauciers* (the cooks responsible for making the sauces) "enlightened chemists," his actual recipes rarely specify quantities of butter or flour or truffles or salt. Instead, they are descriptions of his cooking process: melt the fat, add the meat, listen "for the sound of crackling," pour in the stock, and reduce. It sounds so easy: all you have to do is obey the whims of your senses. This isn't a science experiment, Escoffier seems to be saying, this is hedonism. Let pleasure be your guide.

Escoffier's emphasis on the tongue was the source of his culinary revolution. In his kitchen, a proper cook was a man of exquisite sensitivity, "carefully studying the trifling details of each separate flavor before he sends his masterpiece of culinary art before his patrons." Escoffier's cookbook warns again and again that the experience of the dish — what it actually tastes like — is the only thing that matters: "Experience alone can guide the cook." A chef must be that artist on whom no taste is lost.

But Escoffier knew that he couldn't just serve up some grilled meat and call it a day. His hedonism had to taste haute. After all, he was a chef in the hotels of César Ritz, and his customers expected their food to be worthy of the gilded surroundings and astonishing expense. This is where Escoffier turned to his precious collection of stocks. He used his stocks to ennoble the ordinary sauté, to give the dish a depth and density of flavor. After the meat was cooked in the hot pan (Escoffier preferred a heavy, flat-bottomed *poele*), the meat was taken out to rest, and the dirty pan, full of delicious grease and meat scraps, was deglazed.

Deglazing was the secret of Escoffier's success. The process itself is extremely simple: a piece of meat is cooked at a very high temperature — to produce a nice seared Maillard crust, a cross-linking and caramelizing of amino acids — and then a liquid, such as a rich veal stock, is added.* As the liquid evaporates, it loosens the *fronde*, the burned bits of protein stuck to the bottom of the pan (deglazing also makes life easier for the dishwasher). The dissolved *fronde* is what gives Escoffier's sauces their divine depth; it's what makes *boeuf bourguignon, bourguignon.* A little butter is added for varnish, and, voilà! the sauce is *complet.*

The Secret of Deliciousness

Escoffier's basic technique is still indispensable. Few other cultural forms have survived the twentieth century so intact. Just about every fancy restaurant still serves up variations of his dishes, recycling their bones and vegetable tops into meat stocks. From es-

---

* Escoffier would also use wine, brandy, port, wine vinegar, and — if there was no spare booze lying around — water.

pagnole sauce to sole Véronique, we eat the way he told us to eat. And since what Brillat-Savarin said is probably true — "The discovery of a new dish does more for the happiness of the human race than the discovery of a new star" — it is hard to overestimate Escoffier's importance. Clearly, there is something about his culinary method — about stocks and deglazing and those last-minute swirls of butter — that makes some primal part of us very, very happy.

The place to begin looking for Escoffier's ingenuity is in his cookbooks. The first recipe he gives us is for brown stock (*estouffade*), which he says is "the humble foundation for all that follows." Escoffier begins with the browning of beef and veal bones in the oven. Then, says Escoffier, fry a carrot and an onion in a stockpot. Add cold water, your baked bones, a little pork rind, and a bouquet garni of parsley, thyme, bay leaf, and a clove of garlic. Simmer gently for twelve hours, making sure to keep the water at a constant level. Once the bones have given up their secrets, sauté some meat scraps in hot fat in a saucepan. Deglaze with your bone water and reduce. Repeat. Do it yet again. Then slowly add the remainder of your stock. Carefully skim off the fat (a stock should be virtually fat-free) and simmer for a few more hours. Strain through a fine *chinois*. After a full day of stock-making, you are now ready to *start* cooking.

In Escoffier's labor-intensive recipe, there seems to be little to interest the tongue. After all, everybody knows that the tongue can taste only four flavors: sweet, salty, bitter, and sour. Escoffier's recipe for stock seems to deliberately avoid adding any of these tastes. It contains very little sugar, salt, or acid, and unless one burns the bones (not recommended), there is no bitterness. Why, then, is stock so essential? Why is it the "mother" of Escoffier's cuisine? What do we sense when we eat a profound beef daube, its deglazed bits simmered in stock until the sinewy meat is fit for a spoon? Or, for that matter, when we slurp a bowl of chicken soup, which is just another name for chicken stock? What is it about denatured protein (denaturing is what happens to meat and bones when you cook them Escoffier's way) that we find so inexplicably appealing?

The answer is *umami*, the Japanese word for "delicious." Umami is what you taste when you eat everything from steak to soy sauce. It's what makes stock more than dirty water and deglazing the essential process of French cooking. To be precise, umami is actually the taste of L-glutamate ($C_5H_9NO_4$), the dominant amino acid in the composition of life. L-glutamate is released from life-forms by proteolysis (a shy scientific word for death, rot, and the cooking process). While scientists were still theorizing about the health benefits of tripe, Escoffier was busy learning how we taste food. His genius was getting as much L-glutamate on the plate as possible. The emulsified butter didn't hurt either.

The story of umami begins at about the same time Escoffier invented tournedos Rossini, a filet mignon served with foie gras and sauced with a reduced veal stock and a scattering of black truffles. The year was 1907, and Japanese chemist Kikunae Ikeda asked himself a simple question: What does dashi taste like? Dashi is a classic Japanese broth made from kombu, a dried form of kelp. Since at least A.D. 797, dashi has been used in Japanese cooking the same way Escoffier used stock, as a universal solvent, a base for every dish. But to Ikeda, the dashi his wife cooked for him every night didn't taste like any of the four classic tastes or even like some unique combination of them. It was simply delicious. Or, as the Japanese would say, it was umami.

And so Ikeda began his quixotic quest for this unknown taste. He distilled fields of seaweed, searching for the essence that might trigger the same mysterious sensation as a steaming bowl of seaweed broth. He also explored other cuisines. "There is a taste," Ikeda declared, "which is common to asparagus, tomatoes, cheese and meat but which is not one of the four well-known tastes." Finally, after patient years of lonely chemistry, during which he tried to distill the secret ingredient that dashi and veal stock had in common, Ikeda found his secret molecule. It was glutamic acid, the precursor of L-glutamate. He announced his discovery in the *Journal of the Chemical Society of Tokyo*.

Glutamic acid is itself tasteless. Only when the protein is ionized by cooking, fermentation, or a little ripening in the sun does the molecule degenerate into L-glutamate, an amino acid that the tongue *can* taste. "This study has discovered two facts," Ikeda wrote in his conclusion, "one is that the broth of seaweed contains glutamate and the other that glutamate causes the taste sensation 'umami.'"

But Ikeda still had a problem. Glutamate is an unstable molecule, eager to meld itself to a range of other chemicals, most of which are decidedly *not* delicious. Ikeda knew that he had to bind glutamate to a stable molecule that the tongue did enjoy. His ingenious solution? Salt. After a few years of patient experimentation, Ikeda was able to distill a metallic salt from brown kelp. The chemical acronym of this odorless white powder was MSG, or monosodium glutamate. It was salty, but not like salt. It also wasn't sweet, sour, or bitter. But it sure was delicious.

Ikeda's research, although a seminal finding in the physiology of taste, was completely ignored. Science thought it had the tongue solved. Ever since Democritus hypothesized in the fourth century B.C. that the sensation of taste was an effect of the shape of food particles, the tongue has been seen as a simple muscle. Sweet things, according to Democritus, were "round and large in their atoms," while "the astringently sour is that which is large in its atoms but rough, angular and not spherical." Saltiness was caused by isosceles atoms, while bitterness was "spherical, smooth, scalene and small." Plato believed Democritus, and wrote in *Timaeus* that differences in taste were caused by atoms on the tongue entering the small veins that traveled to the heart. Aristotle, in turn, believed Plato. In *De Anima*, the four primary tastes Aristotle described were the already classic sweet, sour, salty, and bitter.

Over the ensuing millennia, this ancient theory remained largely unquestioned. The tongue was seen as a mechanical organ in which the qualities of foods were impressed upon its papillaed surface. The discovery of taste buds in the nineteenth century gave new credence to this theory. Under a microscope, these cells looked like little keyholes into which our chewed food might fit, thus triggering a taste sensation. By the start of the twentieth century, scientists were beginning to map the tongue, assigning each of the four flavors to a specific area. The tip of the tongue loved sweet things, while the sides preferred sour. The back of the tongue was sensitive to bitter flavors, and saltiness was sensed everywhere. The sensation of taste was that simple.

Unfortunately for Ikeda, there seemed to be no space left on the tongue for his delicious flavor. Umami, these Western scientists said, was an idle theory unique to Japanese food, a silly idea concerned with something called deliciousness, whatever that was. And so while cooks the world over continued to base entire cuisines on dashi, Parmesan cheese, tomato sauce, meat stock, and soy sauce (all potent sources of L-glutamate), science persisted in its naïve and unscientific belief in four, and only four, tastes.

Despite the willful ignorance of science, Ikeda's idea gained a certain cult following. His salty white substance, MSG, a powder that science said couldn't work because we had no means to taste it, nevertheless became an overused staple in everything from cheap Chinese food to bouillon cubes, which used glutamate to simulate the taste of real stock. MSG was even sold in America under the labels Super Seasoning and Accent.* As food products became ever more processed and industrial, adding a dash of MSG became an easy way to create the illusion of flavor. A dish cooked in the microwave tasted as if it had been simmered for hours on the stovetop. Besides, who had time to make meat stock from scratch?

With time, other pioneers began investigating their local cuisines and found their own densities of L-glutamate. Everything from aged cheese to ketchup was rich in this magic little amino acid. Umami even seemed to explain some of the more perplexing idiosyncrasies of the cooking world: why do so many cultures, beginning with ancient Rome, have a fish sauce? (Salted, slightly rotting

---

* MSG is often blamed for the so-called Chinese Restaurant Syndrome, in which exposure to MSG is thought to cause headaches and migraines in certain individuals. But as Jeffrey Steingarten has noted (*It Must've Been Something I Ate*, pp. 85–99), recent research has exonerated both Chinese food and MSG.

anchovies are like glutamate speedballs. They are pure umami.) Why do we dip sushi in soy sauce? (The raw fish, being raw, is low in umami, since its glutamate is not yet unraveled. A touch of soy sauce gives the tongue the burst of umami that we crave.) Umami even explains (although it doesn't excuse) Marmite, the British spread made of yeast extract,* which is just another name for L-glutamate. (Marmite has more than 1750 mg of glutamate per 100 g, giving it a higher concentration of glutamate than any other manufactured product.)

Of course, umami is also the reason that meat — which is nothing but amino acid — tastes so darn good. If cooked properly, the glutamate in meat is converted into its free form and can then be tasted. This also applies to cured meats and cheeses. As a leg of prosciutto ages, the amino acid that increases the most is glutamate. Parmesan, meanwhile, is one of the most concentrated forms of glutamate, weighing in at more than 1200 mg per 100 g. (Only Roquefort cheese has more.) When we add an aged cheese to a pasta, the umami in the cheese somehow exaggerates the umami present elsewhere in the dish. (This is why tomato sauce and Parmesan are such a perfect pair. The cheese makes the tomatoes more tomatolike.) A little umami goes a long way.

And of course, umami also explains Escoffier's genius. The burned bits of meat in the bottom of a pan are unraveled protein, rich in L-glutamate. Dissolved in the stock, which is little more than umami water, these browned scraps fill your mouth with a deep sense of deliciousness, the profound taste of life in a state of decay.

The culture of the kitchen articulated a biological truth of the tongue long before science did because it was forced to feed us. For the ambitious Escoffier, the tongue was a practical problem, and understanding how it worked was a necessary part of creating deli-

---

* Embarrassed food manufacturers often hide the addition of MSG by calling it autolyzed yeast extract on their labels (other pseudonyms for MSG include glutavene, calcium caseinate, and sodium caseinate).

cious dishes. Each dinner menu was a new experiment, a way of empirically verifying his culinary instincts. In his cookbook, he wrote down what every home cook already knew. Protein tastes good, especially when it's been broken apart. Aged cheese isn't just rotten milk. Bones contain flavor. But despite the abundance of experiential evidence, experimental science continued to deny umami's reality. The deliciousness of a stock, said these haughty lab coats, was all in our imagination. The tongue couldn't taste it.

What Ikeda needed before science would believe him was anatomical evidence that we could actually taste glutamate. Anecdotal data from cookbooks, as well as all those people who added fish sauce to their pho, Parmesan to their pasta, and soy sauce to their sushi, wasn't enough.

Finally, more than ninety years after Ikeda first distilled MSG from seaweed, his theory was unequivocally confirmed. Molecular biologists discovered two distinct receptors on the tongue that sense only glutamate and L-amino acids. In honor of Ikeda, they were named the umami receptors. The first receptor was discovered in 2000, when a team of scientists noticed that the tongue contains a modified form of a glutamate receptor already identified in neurons in the brain (glutamate is also a neurotransmitter). The second sighting occurred in 2002, when another umami receptor was identified, this one a derivative of our sweet taste receptors.*

These two separate discoveries of umami receptors on the tongue demonstrated once and for all that umami is not a figment of a hedonist's imagination. We actually have a sense that responds only to veal stock, steak, and dashi. Furthermore, as Ikeda insisted, the tongue uses the taste of umami as its definition of deliciousness. Unlike the tastes of sweet, sour, bitter, and salty, which are sensed

---

* Molecular biology has also revealed how we taste spicy foods. In 2002, researchers discovered that the mouth contains a modified pain receptor — its name is VR1 — which binds capsaicin, the active ingredient in chili peppers. Because the VR1 receptor also detects foods that are hot in temperature, the brain consigns the sensation of excessive heat to any foods that activate our VR1 nerves.

relative to one another (this is why a touch of salt is always added to chocolate, and why melon is gussied up with ham), umami is sensed all by itself. It is that important.

This, of course, is perfectly logical. Why wouldn't we have a specific taste for protein? We love the flavor of denatured protein because, being protein and water ourselves, we need it. Our human body produces more than forty grams of glutamate a day, so we constantly crave an amino acid refill. (Species that are naturally vegetarian find the taste of umami repellent. Unfortunately for vegans, humans are omnivores.) In fact, we are trained from birth to savor umami: breast milk has ten times more glutamate than cow milk. The tongue loves what the body needs.

The Smell of an Idea

Veal stock was not always the glutamate-rich secret of French food. In fact, haute cuisine was not always delicious, or even edible. Before Escoffier began cooking in the new restaurants of the bourgeoisie (unlike his predecessors, he was never a private chef for an aristocrat), fancy cooking was synonymous with ostentation. As long as dinner *looked* decadent, its actual taste was pretty irrelevant. Appearance was everything. Marie-Antoine Carême (1783–1833), the world's first celebrity chef — he cooked for Talleyrand and Czar Alexander I, and he baked Napoleon's wedding cake — epitomized this culinary style. Although Carême is often credited with inventing French cuisine, his food was normally served cold and arranged in epic buffets comprising dozens, sometimes hundreds, of rococo dishes. In Carême's Paris, fancy food was a form of sculpture, and Carême was justifiably famous for his *pièces montées*, which were detailed carvings made of marzipan, pork fat, or spun sugar. Although these sculptures were pretty, they were also inedible. Carême didn't care. "A well-displayed meal," Carême once said, "is improved one hundred percent in my eyes." Such insipid lavishness typified nineteenth-century *service à la française*.

Escoffier considered all this pomp and circumstance ridiculous.

Food was meant to be eaten. He favored *service à la russe* — the Russian style — a system in which the meal was broken down into numbered courses. Unlike Carême's ornate buffets, *service à la russe* featured a single dish per course, which was delivered fresh out of the kitchen. The meal was staggered, and it unfolded in a leisurely culinary narrative: soup was followed by fish, which was followed by meat. And although the chef wrote the menu, the client dictated the tempo and content of the meal. Dessert was the guaranteed happy ending.

This revolution in restaurant service required a parallel revolution in the kitchen. No longer could cooks afford to spend days sculpting marzipan, or molding aspic, or concocting one of Carême's toxically rich stews.* Everything on the menu now had to be *à la minute,* and cooked to order. Flavor had to be manufactured fast. This new level of speed led Escoffier to make his cooking mantra "*Faites simple.*" Every dish, he said, must consist of its necessary ingredients only, and those ingredients must be perfect. A veal stock must contain the very quintessence of veal. Asparagus soup must taste like asparagus, only more so.

There was one added bonus to this new culinary method founded upon simplicity and velocity: food was served *hot.* While Carême feared heat (his lard sculptures tended to melt), Escoffier conditioned his diners to expect a steaming bowl of soup. They wanted their fillets sizzling and the sauce fresh from the deglazed frying pan. In fact, Escoffier's recipes required this efficiency: when meals were served lukewarm, the flavors became disconcertingly one-dimensional. "The customer," Escoffier warned in his cookbook, "finds that the dish is flat and insipid unless it is served absolutely boiling hot."

What Escoffier inadvertently discovered when he started serving food fresh off the stovetop was the importance of our sense of

---

* A typical Carême recipe — *les petits vol-au-vents à la Nesle* — called for two calf udders, twenty cocks' combs and testicles, four whole lamb brains (boiled and chopped), two boned chickens, ten lamb sweetbreads, twenty lobsters, and, just to bind everything together, a few pints of heavy cream.

smell. When food is hot, its molecules are volatile and evaporate into the air. A slowly simmering stock or a clove of garlic sautéed in olive oil can fill an entire kitchen with its alluring odor. Cold food, however, is earth-bound. It relies almost entirely upon the taste buds; the nose is not involved.

But as anyone with a stuffy nose knows, the pleasure of food largely depends on its aroma. In fact, neuroscientists estimate that up to 90 percent of what we perceive as taste is actually smell. The scent of something not only prepares us for eating it (our salivary glands become active), but gives the food a complexity that our five different taste sensations alone can only hint at. If the tongue is the frame for the food — providing us with crucial information about texture, mouth-feel, and the rudiments of taste — the sensations of the nose are what make the food worth framing in the first place.

Escoffier was the first chef to take full advantage of the acutely sensitive nose. Although his food presciently catered to the needs of the tongue (especially its lust for umami), Escoffier aspired to a level of artistry that the tongue couldn't comprehend. As a result, Escoffier's capacious recipes depend entirely upon the flourishes of flavor that we inhale. In fact, all the culinary nuances that so obsessed him — the hint of tarragon in a lobster velouté, the whisper of vanilla in a crème anglaise, the leaf of chervil floating in a carrot soup — are precisely what the unsubtle tongue can't detect. The taste of most flavors is smell.

When we eat, air circulates through the mouth and rises up into the nasal passages, where the gaseous particles of hot food bind to 10 million odor receptors arrayed in an area the size of a thumbprint. When a smell particle binds to the receptor (how they bind no one knows), a surge of ionic energy is created; it travels down the wiry axon, courses through the skull, and connects directly to the brain.

Of course, the receptors in the nose are just the beginning of our sense of smell. The world is an aromatic place. It seems preposterous to expect a nose to have a different receptor for each of the 10,000 to 100,000 discrete odors we can detect. Indeed, our sense of smell — like all of our senses — indulges in a bit of a cellular short-cut, exchanging accuracy for efficiency. And while this might seem stingy, evolution has actually been extremely generous to our sense of smell: odor receptors take up more than 3 percent of the human genome.

Why does the sense of smell require so much DNA? Because our nasal passages are equipped with more than 350 different receptor types, each of which expresses a single receptor gene. These receptor neurons can be activated by many different odorants, including ones that belong to completely distinct chemical families. As a result, before the brain can generate the sensation of any specific scent, the receptors have to work together. They have to transform their scattered bits of information into a coherent representation.

To figure out how this happens, Nobel laureate Richard Axel's lab engineered a fruit fly with a glowing brain, each of its neurons like a little neon light. This was done through the careful insertion of a fluorescent protein in all of the insect's olfactory nerves. But the glow wasn't constant. Axel engineered the fly so that the fluorescent protein turned itself on only when calcium was present in high concentrations inside the cell (active neurons have more calcium). Using some fancy microscopy, Axel's lab group was able to watch — in real time — the patterns of activity within the fly brain whenever it experienced an odor. They could trace the ascent of the smell, how it began as a flicker in a receptor and within milliseconds inflated into a loom of excited cells within the tiny fly nervous system. Furthermore, when the fluorescent fly was exposed to different odors, different areas of its brain lit up. The scent of almonds activated a different electrical grid than the scent of a ripe banana. Axel had found the functional map of smell.

But this imaging of insects, for all of its technical splendor, leaves the real mystery of scent unanswered. Using his neon neurons, Axel can look at the fly's brain and, with shocking accuracy, discern what smell the fly is smelling. He performs this act of mind reading by looking at the fly brain from the *outside*. But how does the *fly* know what it's experiencing? Unless you believe in a little drosophila ghost inside the fly machine, reconstructing its deconstructed smell, this mystery seems impossible to explain. As Axel notes, "No matter how high we get in the fly brain when we map this sensory circuit, the question remains: who in the fly brain is looking down? Who reads the olfactory map? This is our profound and basic problem."

To illustrate the seriousness of this paradox, take an example from our own experience. Imagine you have just inhaled the smell of a deep, dark *demi-glace,* a veal stock that is slowly simmered until it becomes viscous with the gelatin released by its bones. Although Escoffier wanted his sauce to be the distilled essence of veal, he knew that such purity required a lengthy shopping list. You couldn't simply simmer some cow bones in water. The irony behind Escoffier's cooking is that his pursuit of the unadorned dish — a sauce that tasted simply of itself — led him to add a cavalcade of other ingredients, none of which had anything at all to do with the essence of veal. As a result, the *demi-glace* our nose knows is actually composed of many different aromas. Neurons all over the brain light up, reflecting the hodgepodge of smells simultaneously activating our odor receptors. There is the carnal odor of roasted meat; the woodsy scent of the bouquet garni; and the soothing smell of the roux, the flour turning brown with the butter. Underneath those prominent notes, we get the sweetly vegetal waft of the mirepoix, the caramelized tang of tomato paste, and the nutty vapor of evaporated sherry. But how do all of these different ingredients become the specific smell of a *demi-glace,* which tastes like the essence of veal? From the perspective of the nervous system, the challenge is daunting. Within a few milliseconds of being served the *demi-glace,* the mind must bind together the activity of hundreds of distinct smell receptors into a coherent sensation. This is known as the binding problem.

But wait: it gets worse. The binding problem occurs when we experience a sensation that is actually represented as a network of separate neurons distributed across the brain. In the real world, however, reality doesn't trickle in one smell at a time. The brain is constantly confronted with a pandemonium of different odors. As a result, it not only has to bind together its various sensations, it has to decipher which neurons belong to which sensation. For example, that *demi-glace* was probably served as a sauce for a tender fillet of beef, with a side of buttery mashed potatoes. This Escoffier-inspired dish instantly fills the nose with a barrage of distinct scents, from the umber notes of reduced veal stock to the starchy smell of whipped russets. Faced with such a delicious meal, we can either inhale the odor of the dish as a whole — experiencing the overlapping smells as a sort of culinary symphony — or choose to smell each of the items separately. In other words, we can parse our own inputs and, if we so desire, choose to focus on just the smell of potatoes, or *demi-glace,* or a piece of beef served medium rare. Although this act of selective attention seems effortless, neuroscience has no clue how it happens. This is known as the parsing problem.

Parsing and binding are problems because they can't be explained from the bottom up. No matter how detailed our maps of the mind become, the maps still won't explain how a cacophony of cells is bound into the unified perception of a sauce. Or how, at any given moment, we are able to shuttle between our different sensations and separate the smell of the sauce from the smell of the steak. Neuroscience excels at dissecting the bottom of sensation. What our dinner demonstrates is that the mind needs a top.

A Sense of Subjectivity

To make matters even more complicated, what we experience is never limited to our actual sensations. Impressions are always incomplete and require a dash of subjectivity to render them whole. When we bind or parse our sensations, what we are really doing is making judgments about what we *think* we are sensing. This unconscious act of interpretation is largely driven by contextual cues. If you encounter a sensation in an unusual situation — such as the smell of *demi-glace* in a McDonald's — your brain secretly begins altering its sensory verdict. Ambiguous inputs are bound together into a different sensation. The fancy scent of veal stock becomes a Quarter Pounder.

Our sense of smell is particularly vulnerable to this sort of outside influence. Since many odors differ only in their molecular details — and we long ago traded away nasal acuity for better color vision — the brain is often forced to decipher smells based upon *non*-olfactory information. Parmesan cheese and vomit, for example, are both full of butyric acid, which has a pungent top note and a sweetish linger. As a result, blindfolded subjects in experiments will often confuse the two stimuli. In real life, however, such sensory mistakes are extremely rare. Common sense overrules our actual senses.

On an anatomical level, this is because the olfactory bulb is inundated with feedback from higher brain regions. This feedback continually modulates and refines the information garnered by smell receptors. A team of scientists at Oxford has shown that a simple word label can profoundly alter what we think our noses are telling us. When an experimental subject is given odorless air to smell but told he is smelling cheddar cheese, his olfactory areas light up in hungry anticipation. But when that same air arrives with a "body-odor" label, the subject unwittingly shuts down the smell areas of his brain. Although the sensation hasn't changed — it's still just purified air — the mind has completely revised its olfactory response. We unknowingly deceive ourselves.

Escoffier understood this psychological fact. His restaurants were all about the power of suggestion. He insisted that his dishes have fancy names and be served on gilded silver platters. His porcelain was from Limoges, his wineglasses from Austria, and his formidable collection of polished utensils from the estate sales of aristocrats. Escoffier didn't serve steak with gravy; he served *filet de boeuf Richelieu* (which was steak with a well-strained gravy). He made his waiters wear tuxedos, and he helped oversee the rococo decoration of his dining room. A perfect dish, after all, required a perfect mood. Although Escoffier spent eighteen hours a day behind a hot stove, crafting his collection of sauces, he realized that what we taste is ultimately an *idea*, and that our sensations are strongly influenced by their context. "Even horsemeat," Escoffier quipped, "can be delicious when one is in the right circumstances to appreciate it."

This is a suspicious-sounding concept. It smacks of solipsism, relativism, and all those other postmodern -*isms*. But it's our neurological reality. When we sense something, that sensation is immediately analyzed in terms of previous experiences. A *demi-glace* gets filed under sauce, meat, and ways to serve a filet mignon. As the brain figures out what to tell us about this particular *demi-glace*, those previous experiences help us decipher the information being received from the tongue and nose. Is this a good sauce? How does it compare to our memories of other sauces? Do we feel guilty for ordering veal? Was this dish worth the price? Was the waiter rude?

The answers to this cavalcade of unconscious questions determine what we actually experience. Before we've even reached for a second forkful, the *demi-glace* has been ranked and judged, our subjectivity emulsified into our sensation. Thus, what we think we are tasting is only partially about the morsel of matter in the mouth. Equally important is the sum of past experiences enclosed within the brain, for these memories are what *frame* the sensation.

The most persuasive proof of this concept comes from the world of wine. In 2001, Frederic Brochet, of the University of Bordeaux, conducted two separate and very mischievous experiments. In the first test, Brochet invited fifty-seven wine experts and asked them to give their impressions of what looked like two glasses of red and white wine. The wines were actually the same white wine, one of which had been tinted red with food coloring. But that didn't stop the experts from describing the "red" wine in language typically used to describe red wines. One expert praised its "jamminess," while another enjoyed its "crushed red fruit." Not a single one noticed it was actually a white wine.

The second test Brochet conducted was even more damning. He took a middling Bordeaux and served it in two different bottles. One bottle was labeled as a fancy Grand Cru. The other bottle was labeled as an ordinary *vin du table*. Despite the fact that they were served the exact same wine, the experts gave the differently labeled bottles nearly opposite ratings. The Grand Cru was "agreeable, woody, complex, balanced, and rounded," while the *vin du table* was "weak, short, light, flat, and faulty." Forty experts said the wine with the fancy label was worth drinking, while only twelve said the cheap wine was.

What these wine experiments illuminate is the omnipresence of subjectivity. When we take a sip of wine, we don't taste the wine first and the cheapness or redness second. We taste everything all at once, in a single gulp of *this-wine-is-red,* or *this-wine-is-expensive.* As a result, the wine experts sincerely believed that the white wine was red, and that the cheap wine was expensive. And while they were pitifully mistaken, the mistakes weren't entirely their fault. Our human brain has been designed to believe itself, wired so that prejudices feel like facts, opinions are indistinguishable from the actual sensation. If we *think* a wine is cheap, it will taste cheap. And if we think we are tasting a Grand Cru, then we will taste a Grand Cru. Our senses are vague in their instructions, and we parse their suggestions based on whatever other knowledge we can summon to the surface. As Brochet himself notes, our expectations of what the wine will taste like "can be much more powerful in determining how you taste a wine than the actual physical qualities of the wine itself."

The fallibility of our senses — their susceptibility to our mental biases and beliefs — poses a special problem for neural reductionism. The taste of a wine, like the taste of everything, is not merely the sum of our inputs and cannot be solved in a bottom-up fashion. It cannot be deduced by beginning with our simplest sensations and extrapolating upward. This is because what we experience is not what we sense. Rather, experience is what happens when sensations are interpreted by the subjective brain, which brings to the moment its entire library of personal memories and idiosyncratic desires.* As the philosopher Donald Davidson argued, it is ulti-

---

* The importance of top-down feedback is also apparent *outside* of the brain. Take, for example, the taste of a well-seared steak. Escoffier always believed that a steak seared at high

mately impossible to distinguish between a subjective contribution to knowledge that comes from our selves (what he calls our "scheme") and an objective contribution that comes from the outside world ("the content"). In Davidson's influential epistemology, the "organizing system and something waiting to be organized" are hopelessly interdependent. Without our subjectivity we could never decipher our sensations, and without our sensations we would have nothing about which to be subjective. Before you can taste the wine you have to judge it.

But even if we could — by some miracle of Robert Parkeresque objectivity — taste the wine *as it is* (without the distortions of scheming subjectivity), we would still all experience a different wine. Science has long known that our sensitivity to certain smells and tastes varies by as much as 1,000 percent between individuals. On a cellular level, this is because the human olfactory cortex, the part of the brain that interprets information from the tongue and nose, is extremely plastic, free to arrange itself around the content of our individual experiences. Long after our other senses have settled down, our senses of taste and smell remain in total neural flux. Nature designed us this way: the olfactory bulb is full of new neurons. Fresh cells are constantly being born, and the survival of these cells depends upon their activity. Only cells that respond to the smells and tastes we are actually exposed to survive. Everything else withers away. The end result is that our brains begin to reflect what we eat.

---

temperature is juicier, for the seared crust "seals in the meat's natural juices." This is completely false. (Even Escoffier made mistakes.) Technically speaking, a steak cooked at high temperatures contains *less* of its own juice, as that alluring sizzling noise is actually the sound of the meat's own liquid evaporating into thin air. (For maximum retention of natural juices, cook the steak slow and steady, and don't salt until the end.) Nevertheless, what Escoffier noticed is true: even if a well-seared steak is literally drier, it still *tastes* juicier. The disquieting explanation of this culinary illusion is that a well-seared steak — its Maillard crust crisp and crackling, its interior plush and bloody — makes us drool in anticipation. As a result, when we eat the more appetizing — yet less juicy — steak, the meat *seems* to be juicier. However, what we are actually sensing is our own saliva, which the brain induced the salivary glands to release. Our personal decision to drool warps the sensory experience of the steak.

The best documented example of this peripheral plasticity concerns the chemical androstenone, a steroid that occurs in urine and sweat and has been proposed as a human pheromone. When it comes to smelling androstenone, humans fall into three separate categories. The first group simply can't smell it. The second group is made up of very sensitive smellers who can detect fewer than ten parts per trillion of androstenone and who find the odor extremely unpleasant (it smells like urine to them). And the third group consists of smellers who are less sensitive to the odor but perceive it in oddly pleasant ways, such as sweet, musky, or perfume-like. What makes these differences in sensory experience even more interesting is that experience modulates sensitivity. Subjects repeatedly exposed to androstenone become more sensitive to it, thanks to feedback from the brain. This feedback causes the stem cells in nasal passages to create more androstenone-sensitive odor receptors. The new abundance of cells alters the sensory experience. What was once a perfume becomes piss.

Of course, in the real world (as opposed to the laboratory) it is we who control our experiences. We choose what to eat for dinner. Escoffier understood this first: he wanted his customers to order their own meals precisely because he never knew what they might want. Would they order a beef stew or some salmon quenelles? A bowl of bouillon or some sweetbreads larded with truffles? While Escoffier made his customers obey a few basic rules (no white wine with beef, no smoking between courses, no blanquette after a creamy soup, and so forth), he realized that every diner had a separate set of idiosyncratic desires. This is why he invented the menu: so his guests could choose dinner for themselves.

And every time a customer devoured one of Escoffier's dishes, choosing the fillet over the *rouget,* the sensations of that person's tongue were altered. When Escoffier was working at the Savoy in London (a joint venture with César Ritz), he had faith that he could educate even the palates of the British. At first, Escoffier was horrified by how his new patrons defied his carefully arranged menu. (He refused to learn English out of fear, he later said, that he would come to cook like the English.) Some customers would order two creamy dishes (a real faux pas), while others would want meat without a sauce, or would only have a little soup for supper. To teach his British guests how to eat a meal properly, Escoffier decided that any party larger than four people dining in his London restaurant could only have whatever dishes he put in front of them. He invented the chef's tasting menu as an educational tool, for he was confident that people could *learn* how to eat. Over time, the English could become more French. He was right: because the sense of taste is extremely plastic, it can be remodeled by new experiences. It's never too late to become a gourmet.

Since the publication of his encyclopedic cookbook in 1903, Escoffier's culinary inventions have gone on to modify untold numbers of olfactory cortices, noses, and tongues. His recipes have literally changed our sensations, teaching us what to want and how our most coveted dishes should be served. This is the power of good cooking: it invents a new kind of desire. From Escoffier, we learned how to love flavorful stocks, viscous sauces, and all the silver accoutrements of fancy French food. He led us to expect food to taste like its essence, no matter how many extra ingredients that required. And while his love of butter may have shortened our own life spans, the wisdom of his recipes has made our short lives a little bit happier.

How did Escoffier invent such a potent collection of dishes? By taking his own experiences seriously. He knew that deliciousness was deeply personal, and that any analysis of taste must begin with the first-person perspective. Like Ikeda, he listened not to the science of his time, which treated the tongue like a stranger, but to the diversity of our cravings and the whims of our wants. Our pleasure was his experimental guide. As Escoffier warned at the start of his cookbook, "No theory, no formula, and no recipe can take the place of experience."

Of course, the individuality of our experiences is what science will never be able to solve. The fact is, each of us literally inhabits a different brain, tuned to the tenor of our private desires. These desires have been molded — at the level of our neurons — by a lifetime of eating. Escoffier's *Guide Culinaire* is more than six hundred pages long because he knew that there was no single recipe, no matter how much umami and cream it contained, that would satisfy everyone. The individuality of taste, which is, in a way, the only aspect of taste that really matters, cannot be explained by science. The subjective experience is irreducible. Cooking is a science *and* an art. As the chef Mario Batali once said about one of his recipes, "If it works, it is true."

Chapter 4

Marcel Proust

The Method of Memory

Even a bureau crammed with souvenirs,
Old bills, love letters, photographs, receipts,
Court depositions, locks of hair in plaits,
Hides fewer secrets than my brain could yield.
It's like a tomb, a corpse-filled Potter's field,
A pyramid where the dead lie down by scores.
I am a graveyard that the moon abhors.

— Charles Baudelaire, LXXVI

THE TITLE OF Marcel Proust's novel *In Search of Lost Time* is literal.* In his fiction, Proust was searching for the hidden space where time stops. Obsessed with "the incurable imperfection in the very essence of the present moment," Proust felt the hours flowing over him like cold water. Everything was ebbing away. A sickly thirty-something, Proust had done nothing with his life so far except

---

* Proust's epic *A la Recherche du Temps Perdu* has actually been given two different titles in English. The first title, *Remembrance of Things Past,* was given to the novels by their translator C. K. Scott Moncrieff. It is not a literal translation (Moncrieff borrowed it from a Shakespeare sonnet). While this title effectively evokes the content of Proust's novel, it fails to capture either Proust's obsession with time or the fact that his fiction was a search for something. Proust himself took the title very seriously, considering everything from *The Stalactites of the Past* to *Reflections in the Patina* to *Lingered Over Days* to *Visit from a Past That Lingers On.* After a few months of considering these options, Proust settled on *A la Recherche du Temps Perdu.* In 1992, the translator D. J. Enright gave Proust's novel the far more literal English title *In Search of Lost Time.*

accumulate symptoms and send self-pitying letters to his mother. He wasn't ready to die.

And so, seeking a taste of immortality, Proust became a novelist. Deprived of a real life — his asthma confined him to his bedroom — Proust made art out of the only thing he had: his memory. Nostalgia became his balm, "for if our life is vagabond, our memory is sedentary." Proust knew that every time he lost himself in a recollection he also lost track of time, the tick-tock of the clock drowned out by the echoey murmurs of his mind. It was there, in his own memory, that he would live forever. His past would become a masterpiece.

Emboldened by this revelation, Proust began writing. And writing. And writing. He disappeared into his drafts, emerging only, he said, "when I need help remembering." Proust used his intuition, his slavish devotion to himself and his art, to refine his faith in memory into an entire treatise. In the stuffy silence of his Parisian studio, he listened so intently to his sentimental brain that he discovered how it operated.

What sort of truth did Proust discover? It's a cliché to say that he described a very real milieu, a snapshot of Parisian society during the glory days of glamour. Other literary scholars focus on the style of his sentences, their rapturous roll and lulling cadences as he describes yet another dinner party. Proust covers vast distances within the space of periods (one sentence is 356 words long), and often begins with the obscure detail (the texture of a napkin or the noise of water in the pipes) and ends with an inductive meditation on all things. Henry James, no slouch at verbosity himself, defined Proust's style as "an inconceivable boredom associated with the most extreme ecstasy which it is possible to imagine."

But all those beliefs in Proust's panache and artistic skill, while true, ignore the seriousness of his thoughts on memory. Although he had a weak spot for subclauses and patisserie, somehow, by sheer force of adjectives and loneliness, he intuited some of modern neuroscience's most basic tenets. As scientists dissect our remembrances into a list of molecules and brain regions, they fail to realize

A portrait of Marcel Proust by Jacques Emile Blanche,
completed in 1892

that they are channeling a reclusive French novelist. Proust may not have lived forever, but his theory of memory endures.

Intuitions

Proust wouldn't be surprised by his prophetic powers. He believed that while art and science both dealt in facts ("The impression is for the writer what experimentation is for the scientist"), only the artist was able to describe reality as it was actually experienced. Proust was confident that every reader who read his novel would "recognize in his own self what the book says . . . This will be the proof of its veracity."

Proust learned to believe in the strange power of art from the philosopher Henri Bergson.* When Proust began writing the *Search*, Bergson was becoming a celebrity. The metaphysician sold out opera halls, the intellectual tourists listening with rapt attention to his discussions of *élan-vital*, comedy, and "creative evolution."† The essence of Bergson's philosophy was a fierce resistance to a mechanistic view of the universe. The laws of science were fine for inert matter, Bergson said, for discerning the relationships between atoms and cells, but us? We had a consciousness, a memory, a being. According to Bergson, this reality — the reality of our self-consciousness — could not be reduced or experimentally dissected. He believed that we could only understand ourselves through *intuition*, a process that required lots of introspection, lazy days contemplating our inner connections. Basically, it was bourgeois meditation.

Proust was one of the first artists to internalize Bergson's philosophy. His literature became a celebration of intuition, of all the truths we can know just by lying in bed and quietly thinking. And while Bergson's influence was not without its anxiety for Proust — "I have enough to do," he wrote in a letter, "without trying to turn the philosophy of M. Bergson into a novel!" — Proust still couldn't resist Bergsonian themes. In fact, Proust's thorough absorption of Bergson's philosophy led him to conclude that the nineteenth-century novel, with its privileging of things over thoughts, had everything exactly backward. "The kind of literature which contents itself with 'describing things,'" Proust wrote, "with giving them merely a mis-

---

* Proust attended Bergson's lectures given at the Sorbonne from 1891 to 1893. In addition, he read Bergson's *Matter and Memory* in 1909, just as he was beginning to compose *Swann's Way*. In 1892, Bergson married Proust's cousin. However, there is only one recorded conversation between Proust and Bergson, in which they discussed the nature of sleep. This conversation gets retold in *Sodom and Gomorrah*. To the philosopher, however, Proust would remain nothing more than the cousin who had bought him an excellent box of earplugs.

† His appearance at Columbia University in 1913 caused the first traffic jam ever in New York City.

erable abstract of lines and surfaces, is in fact, though it calls itself realist, the furthest removed from reality." As Bergson insisted, reality is best understood *subjectively,* its truths accessed intuitively.

But how could a work of fiction demonstrate the power of intuition? How could a novel prove that reality was, as Bergson put it, "ultimately spiritual, and not physical"? Proust's solution arrived in the unexpected form of a buttery cookie flavored with lemon zest and shaped like a seashell. Here was a bit of matter that revealed "the structure of his spirit," a dessert that could be "reduced back into its psychological elements." This is how the *Search* begins: with the famous madeleine, out of which an entire mind unfolds:

> No sooner had the warm liquid mixed with the crumbs touched my palate than a shudder ran through me and I stopped, intent upon the extraordinary thing that was happening to me. An exquisite pleasure had invaded my senses, something isolated, detached, with no suggestion of its origin. And at once the vicissitudes of life had become indifferent to me, its disasters innocuous, its brevity illusory; it was me. I had ceased to feel mediocre, contingent, mortal. Whence could it have come to me, this all-powerful joy? I sensed that it was connected with the taste of the tea and the cake, but that it infinitely transcended those savours, could not, indeed, be of the same nature. Whence did it come? What did it mean? How could I seize it and apprehend it?
>
> I drank a second mouthful, in which I find nothing more than in the first, then a third, which gives me rather less than the second. It is time to stop; the potion is losing its magic. It is plain that the truth I am seeking lies not in the cup but in myself.

This gorgeous paragraph captures the essence of Proust's art, the truth wafting up like steam from a limpid cup of tea. And while the madeleine was the trigger for Proust's epiphany, this passage isn't about the madeleine. The cookie is merely a convenient excuse for Proust to explore his favorite subject: himself.

What did Proust learn from these prophetic crumbs of sugar, flour, and butter? He actually intuited a lot about the structure of our brain. In 1911, the year of the madeleine, physiologists had no idea how the senses connected inside the skull. One of Proust's deep insights was that our senses of smell and taste bear a unique burden of memory:

> When from a long distant past nothing subsists, after the people are dead, after the things are broken and scattered, *taste and smell alone,* more fragile but enduring, more unsubstantial, more persistent, more faithful, remain poised a long time, like souls, remembering, waiting, hoping, amid the ruins of all the rest; and bear unflinchingly, in the tiny and almost impalpable drop of their essence, the vast structure of recollection.

Neuroscience now knows that Proust was right. Rachel Herz, a psychologist at Brown, has shown — in a science paper wittily entitled "Testing the Proustian Hypothesis" — that our senses of smell and taste are uniquely sentimental. This is because smell and taste are the only senses that connect directly to the hippocampus, the center of the brain's long-term memory. Their mark is indelible. All our other senses (sight, touch, and hearing) are first processed by the thalamus, the source of language and the front door to consciousness. As a result, these senses are much less efficient at summoning up our past.

Proust intuited this anatomy. He used the taste of the madeleine and the smell of the tea to channel his childhood.* Just looking at the scalloped cookie brought back nothing. Proust even goes so far as to blame his sense of sight for obscuring his childhood memories in the first place. "Perhaps because I had so often seen such madeleines without tasting them," Proust writes, "their image had disas-

---

* A. J. Liebling, the celebrated hedonist and *New Yorker* writer, once wrote: "In the light of what Proust wrote with so mild a stimulus (the quantity of brandy in a madeleine would not furnish a gnat with an alcohol rub), it is the world's loss that he did not have a heartier appetite."

  Liebling would be happy to know that Proust actually had an excellent appetite. Though he only ate one meal a day (doctor's orders), Proust's dinner was Lieblingesque. A sample menu included two eggs in cream sauce, three croissants, half a roast chicken, French fries, grapes, beer, and a few sips of coffee.

sociated itself from those Combray days." Luckily for literature, Proust decided to put the cookie in his mouth.

Of course, once Proust began to remember his past, he lost all interest in the taste of the madeleine. Instead, he became obsessed with how he *felt* about the cookie, with what the cookie *meant* to him. What else would these crumbs teach him about his past? What other memories could emerge from these magic mouthfuls of flour and butter?

In this Proustian vision, the cookie is worthy of philosophy because in the mind, everything is connected. As a result, a madeleine can easily become a revelation. And while some of Proust's ensuing mental associations are logical (for example, the taste of the madeleine and the memory of Combray), others feel oddly random. Why does the cookie also bring to his mind "the game wherein the Japanese amuse themselves by filling a porcelain bowl with water and steeping in it little pieces of paper"? And why does a starchy napkin remind him of the Atlantic Ocean, which "swells in blue and bosomy undulations"? An honest chronicler of his own brain, Proust embraced such strange associations precisely because he couldn't explain them. He understood that idiosyncrasy was the essence of personality. Only by meticulously retracing the loom of our neural connections — however nonsensical those connections may be — can we understand ourselves, *for we are our loom.* Proust gleaned all of this wisdom from an afternoon tea.

The Lie of Yesterday

So there is time, and there is memory. Proust's fiction, which is mostly nonfiction, explores how time mutates memory. Just before Marcel takes a sip of his lime-flower tea, he issues a bleak warning to his reader: "It is a labor in vain to attempt to recapture memory: all the efforts of our intellect must prove futile . . ." Why does Proust think the past is so elusive? Why is the act of remembering a "labor in vain"?

These questions cut to the core of Proust's theory of memory. Simply put, he believed that our recollections were phony. Although they felt real, they were actually elaborate fabrications. Take the madeleine. Proust realized that the moment we finish eating the cookie, leaving behind a collection of crumbs on a porcelain plate, we begin warping the memory of the cookie to fit our own personal narrative. We bend the facts to suit our story, as "our intelligence reworks the experience." Proust warns us to treat the reality of our memories carefully, and with a degree of skepticism.

Even within the text itself, the Proustian narrator is constantly altering his remembered descriptions of things and people, particularly his lover Albertine. Over the course of the novel, Albertine's beauty mark migrates from her chin to her lip to a bit of cheekbone just below her eye. In any other novel, such sloppiness would be considered a mistake. But in the *Search,* the instability and inaccuracy of memory is the moral. Proust wants us to know that we will never know where Albertine's beauty mark really is. "I am obliged to depict errors," Proust wrote in a letter to Jacques Rivière, "without feeling compelled to say that I consider them to be errors." Because *every* memory is full of errors, there's no need to keep track.

The strange twist in the story is that science is discovering the molecular truth behind these Proustian theories. Memory *is* fallible. Our remembrance of things past is imperfect.

The dishonesty of memory was first scientifically documented by Freud, by accident. In the course of his psychotherapy, he dealt with a staggering number of women who traced their nervous hysterias back to sexual abuse in their childhood. To explain their confessions, Freud was forced to confront two equally dismaying scenarios. Either the women were lying, or sexual molestation was disturbingly common in bourgeois Vienna. In the end, Freud realized that the real answer was beyond the reach of his clinic. The psychotherapist would never discover what really happened, for the moment the women "remembered" their sexual abuse, they also created sincere memories. Even if their tales of abuse were fabrications, the women weren't technically lying, since they believed every word of it. Our recollections are cynical things, designed by the brain to always *feel* true, regardless of whether or not they actually occurred.

For most of the twentieth century, neuroscience followed Freud's pose of indifference. It wasn't interested in investigating the fictionality of memory, or how the act of remembering might alter a memory. Scientists assumed that memories are just shelved away in the brain, like dusty old books in a library. But this naïve approach eventually exhausted itself. In order to investigate the reality of our past, in order to understand memory as we actually experience it, scientists needed to confront the specter of memory's lie.

Every memory begins as a changed connection between two neurons. This fact was first intuited by Santiago Ramon y Cajal, who won the Nobel Prize for Medicine in 1906. Cajal's scientific process was simple: he stared at thin slices of brain under a microscope and let his imagination run wild. (Cajal called his science a "speculative cavort.")* At the time, scientists assumed that the human brain's neurons were connected in a seamless reticular web, like electrical wires linked in a circuit. Cajal, however, believed that every neuron was actually an island, entirely bounded by its own membrane (an idea that wasn't confirmed until electron microscopy studies in the 1950s). But if neurons don't touch, then how do they form memories and exchange information? Cajal hypothesized that the vacant gaps between cells — what we now call synaptic clefts — were the secret sites of communication. What Joseph Conrad said about maps is also true of the brain: the most interesting places are the empty spaces, for they are what will change.

Cajal was right. Our memories exist as subtle shifts in the strength of synapses, which make it easier for neurons to communicate with one another. The end result is that when Proust tastes a madeleine,

* In his *Advice for a Young Investigator,* Cajal wrote, "No one without a certain intuition — a divinatory instinct for perceiving the idea behind the fact and the law behind the phenomenon — will devise a reasonable solution, whatever his gifts as an observer."

the neurons downstream of the cookie's taste, the ones that code for Combray and Aunt Leonie, light up. The cells have become inextricably entwined; a memory has been made. While neuroscientists still don't know how this happens,* they do know that the memory-making process needs new proteins. This makes sense: proteins are the bricks and mortar of life, and a remembrance requires some cellular construction. The moment in time is incorporated into the architecture of the brain.

But in a set of extraordinary experiments done at NYU in 2000 by Karim Nader, Glenn Shafe, and Joseph LeDoux, scientists demonstrated that *the act of remembering* also changes you. They proved this by conditioning rats to associate a loud noise with a mild electrical shock. (When it comes to pain, the mind is a quick learner.) As predicted, injecting a chemical that stops new proteins from being created also prevented the rats from creating a fearful memory. Since their brains were unable to connect their context to the electrical shock, the shock was always shocking.

But Nader, LeDoux, and Shafe took this simple experiment one step further. First, they made sure that the rats had a strong memory associating the shock with the noise. They wanted rodents that would cower in fear whenever the sound was played. After letting this memory solidify for up to forty-five days, they re-exposed the rats to the scary noise and injected a protein inhibitor into their brains. But what made their experiment different was its timing. Instead of interrupting the process of making a memory, they interrupted the process of *remembering* a memory, injecting the noxious chemical at the exact moment the rats were recalling what the noise meant. According to the dogma of remembrance, nothing much should have happened. The long-term memory should exist independently of its recall, filed away in one of the brain's protected file cabinets. After the poison is flushed out of their cells, the rats

---

* The likely suspects include an increased density of neurotransmitter receptors; a greater release of neurotransmitter with every excitatory event; some kind of retrograde messenger, like nitrous oxide; or some combination of all of the above.

should remember their fear. The noise should still remind them of the shock.

But this isn't what happened. When Nader and his group blocked the rats from *remembering* their fearful memory, the original memory trace also disappeared. After only a single interruption of the recollection process, their fear was erased. The rats became amnesiacs.*

At first glance, this experimental observation seems incongruous. After all, we like to think of our memories as being immutable impressions, somehow separate from the act of remembering them. But they aren't. A memory is only as real as the last time you remembered it. The more you remember something, the less accurate the memory becomes.

The Nader experiment, simple as it seems, requires science to completely re-imagine its theories of remembering. It reveals memory as a ceaseless process, not a repository of inert information. It shows us that every time we remember anything, the neuronal structure of the memory is delicately transformed, a process called reconsolidation. (Freud called this process *Nachtraglichkeit*, or "retroactivity.") The memory is altered in the absence of the original stimulus, becoming less about what you remember and more about you. So the purely objective memory, the one "true" to the original taste of the madeleine, is the one memory you will never know. The moment you remember the cookie's taste is the same moment you forget what it really tasted like.

Proust presciently anticipated the discovery of memory reconsolidation. For him, memories were like sentences: they were things you never stopped changing. As a result, Proust was not only an avid sentimentalist, he was also an insufferable rewriter. He scribbled in the margins of his drafts and then, when the margins over-

* Neuroscientists are now looking at reconsolidation as a possible treatment for post-traumatic stress disorder (PTSD) and drug addiction. By blocking destructive memories as they are being recalled, scientists hope to erase the anxieties and addictions entirely.

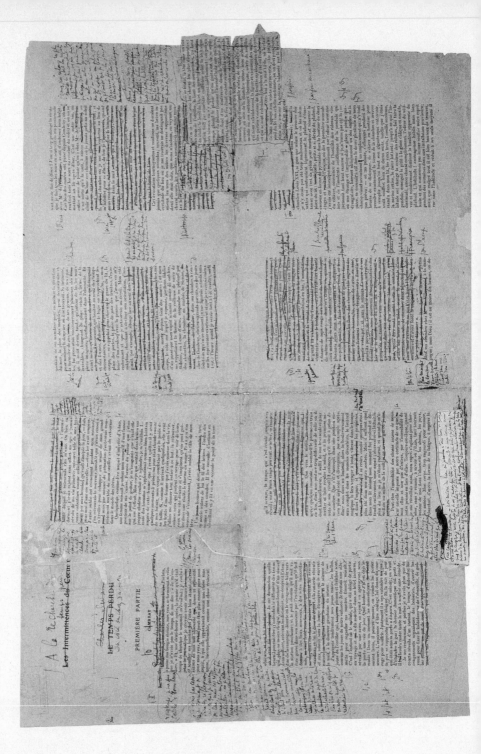

flowed, he supplemented his pages with *paperoles,* little cut pieces of paper that he would paste onto his original manuscript. Nothing he wrote was ever permanent. It was not uncommon for him to stop the printing presses at his own expense.

Clearly, Proust believed in the writing *process.* He never outlined his stories first. He thought that the novel, like the memories it unfaithfully described, must unfurl naturally. While the *Search* began as an essay against the literary critic Charles Augustin Sainte-Beuve — Proust argued that literature *cannot* be interpreted in terms of the literal life of the artist — it quickly swelled into an epic about childhood, love, jealousy, homosexuality, and time. Then World War I intervened, the printing presses were turned into tanks, and Proust's novel, having no commercial outlet, metastasized from a formidable half a million words into a Talmudic 1¼ million words. At the same time, the love of Proust's life, Alfred Agnostelli, tragically crashed his plane into the sea. Proust lavished his grief on a whole new plot line in which the character Albertine, Alfred's doppelganger in the novel, also dies.

For a novel about memory, the plasticity of the novel's narrative was one of its most realistic elements. Proust was always refining his fictional sentences in light of new knowledge, altering his past words to reflect his present circumstances. On the last night of his life, as he lay prostrate in bed, weakened by his diet of ice cream, beer, and barbiturates, he summoned Celeste, his beloved maid, to take a little dictation. He wanted to change a section in his novel that described the slow death of a character, since he now knew a little bit more about what dying was like.

The uncomfortable reality is that we remember in the same way that Proust wrote. As long as we have memories to recall, the margins of those memories are being modified to fit what we know *now.* Synapses are crossed out, dendrites are tweaked, and the memorized moment that feels so honest is thoroughly revised. In his

OPPOSITE: *Page proof for the* Search. *The book had already been sent to the printer, but Proust insisted on making extensive changes.*

own lifetime, Proust never saw the complete *Search* printed. For him, the work would always remain malleable, just like a memory.

Before Nader created his forgetful rats in 2002, neuroscientists had avoided the murky area of remembrance and reconsolidation. Instead, scientists focused on meticulously outlining the molecules responsible for *storing* a memory. They assumed that a memory was like a photograph, a fixed snapshot of a moment, so it didn't really matter how the memory was actually remembered. If only they had read Proust.

One of the morals of the *Search* is that every memory is inseparable from the moment of its recollection. This is why Proust devoted fifty-eight tedious pages to the mental state of the narrator *before* he ate a single madeleine. He wanted to show how his current condition distorted his sense of the past. After all, when Marcel was actually a child in Combray, eating madeleines to his heart's content, all he wanted was to escape his small town. But once he escaped, Marcel incessantly dreamed of recovering the precious childhood that he had so wantonly squandered. This is the irony of Proustian nostalgia: it remembers things as being far better than they actually were. But Proust, at least, was acutely aware of his own fraudulence. He knew that the Combray he yearned for was not the Combray that was. (As Proust put it, "The only paradise is paradise lost.") This wasn't his fault: there simply is no way to describe the past without lying. Our memories are not *like* fiction. They *are* fiction.

Proust's novels tantalizingly toy with the fictionality of memory in a very postmodern way: the narrator, who identifies himself as Marcel only once in three thousand pages,* begins sentences with *I.* Like Proust, the narrator has translated Ruskin, dabbled in high society's parlors, and is now a sickly recluse writing *In Search of Lost Time.* And some characters, though Proust denied it to the bitter end, are thinly veiled acquaintances. In his books, fiction and reality

---

* Here's the line that confirms Proust is the narrator: "'My —' or 'My darling' followed by my Christian name which, if we give the narrator the same name as the author of this book, would be 'My Marcel,' or 'My darling Marcel.'"

are hopelessly intertwined. But Proust, always coy, denied this veri-similitude:

> In this book, in which every fact is fictional and in which not a single character has been based on a living person, in which everything has been invented by me according to the needs of my demonstration, I must state to the credit of my country that only Françoise's million-aire relatives, who interrupted their retirement in order to help their needy niece, are real people, existing in the world.

This passage comes toward the end of *Time Regained,* the last book of the *Search.* It is not a denial of the novel's mirroring of real-ity so much as an attempt to explode any investigation of it. Proust gives a sarcastic point of intersection (Françoise's millionaire rela-tives) as the sole meeting place of reality and literature, truth and memory. Proust here is being more than a little disingenuous. The novel and the life, the journalist and the fabulist, are really hope-lessly blurred together. Proust likes it that way because that's how memory actually is. As he warned at the end of *Swann's Way,* "How paradoxical it is to seek in reality for the pictures that are stored in one's memory . . . The memory of a particular image is but regret for a particular moment; and houses, roads, avenues are as fugitive, alas, as the years."

In this Proustian paradigm, memories do not directly represent reality. Instead, they are imperfect copies of what actually hap-pened, a Xerox of a Xerox of a mimeograph of the original photo-graph. Proust intuitively knew that our memories required this transformative process. If you prevent the memory from changing, it ceases to exist. Combray is lost. This is Proust's guilty secret: we have to misremember something in order to remember it.

Sentimental Proteins

Some memories exist outside time, like magic carpets folded deli-cately in our mind. Unconscious recollection is at the heart of Proust's model of memory because even as our memories define us, they seem to exist without us. When *Swann's Way* begins, Proust has forgotten all about the sugary pastries of his childhood. Combray is just another Parisian suburb. But then, when he eats the madeleine that reminds him of Aunt Leonie, and the scent of the tea conspires with the texture of the napkin, the memory returns to haunt him, like a ghost. Lost time is found. Proust worshipped these sudden epiphanies of the past because they seemed more truthful, less corrupted by the lies of the remembering process. Marcel is like the boy described by Freud who liked to lose his toys because he so loved to find them.

But how do these unconscious memories persist? And how do we remember them after they have already been forgotten? How does an entire novel, or six of them, just hide away in the brain, waiting patiently for a madeleine?

Until a few years ago, neuroscience had no explanation for Proust's *moments bienheureux* ("fortunate moments"), those shattering epiphanies when recollection appears like an apparition. The standard scientific model for memory revolved around enzymes and genes that required lots of reinforcement in order to be activated. The poor animals used for these experiments had to be trained again and again, their neurons bullied into altering their synaptic connections. Senseless repetition seemed to be the secret of memory.

Unfortunately for neuroscience, this isn't the way most memories are made. Life only happens once. When Proust remembers the madeleine in *Swann's Way,* it wasn't because he'd eaten lots of madeleines. In fact, the opposite was true. Proust's memory is hauntingly specific and completely unexpected. His memory of Combray, cued by some chance crumbs, interrupts his life, intruding for no logical reason, "with no suggestion of its origin." Proust is shocked by his past.

These literary memories are precisely the sort of remembrances that the old scientific models couldn't explain. Those models don't seem to encapsulate the randomness and weirdness of the memory we live in. They don't describe its totality, the way memories appear and disappear, the way they change and float, sink and swell. Our memories obsess us precisely because they disobey every logic, because we never know what we will retain and what we will forget.

But what makes science so wonderful is its propensity to fix itself. Like Proust, who was perfecting sentences until the printer set his type, scientists are never satisfied with their current version of things. In the latest draft of the science of memory, the theorizing has undergone a remarkable plot twist. Scientific rumors are emerging that may unlock the molecular details of how our memories endure even when we've forgotten all about them.

This theory, published in 2003 in the journal *Cell,* remains controversial. Nevertheless, the elegance of its logic is tantalizing. Dr. Kausik Si, a former postdoc in the lab of Nobel laureate Eric Kandel, believes he has found the "synaptic mark" of memory, the potent grain that persists in the far electrical reaches of neurons.* The molecule he and Dr. Kandel have discovered could very well be the solution to Proust's search for the origin of the past.

Si began his scientific search by trying to answer the question posed by the madeleine. How do memories last? How do they escape the withering acids of time? After all, the cells of the brain, like all cells, are in constant flux. The average half-life of a brain protein is only fourteen days. A small subset of our hippocampal neurons dies and is reborn; the mind is in a constant state of reincarnation. And yet Si knew that the past feels immutable. Si concluded that our memories must be made of a very strong material, something sturdier even than our cells.

But a neuronal memory cannot simply be strong: it must also be specific. While each neuron has only a single nucleus, it has a teeming mass of dendritic branches. These twigs wander off in every direction, connecting to other neurons at dendritic synapses (imagine two trees whose branches touch in a dense forest). It is at these tiny crossings that our memories are made: not in the trunk of the neuronal tree, but in its sprawling canopy.

How does a cell alter a remote part of itself? Si realized that none

* Full disclosure: I worked for several years for Dr. Si.

of the conventional models of memory could explain such a phenomenon.* There must be something else, some unknown ingredient, which *marked* a specific branch as a memory. The million dollar-question was, What molecule did the marking? What molecular secret lurked in our dendritic densities, silently waiting for a cookie?

Si began his search by thinking through the problem. He knew that any synaptic marker would have to be able to turn on messenger RNA (mRNA), since mRNA helps make proteins, and new memories need new proteins. Furthermore, because mRNA is regulated where memories are regulated — in the dendrites — activating mRNA would allow a neuron to selectively modify its details. This insight led Si to frog eggs. He had heard of a molecule that was able to activate specific scraps of mRNA during the egg's development. This same molecule also happened to be present in the hippocampus, the brain's memory center. Its ignominious name was CPEB, for cyptoplasmic polyadenylation element binding protein.

To see if CPEB was actually important for memory (and not just for frog zygotes), Si began by searching for it in purple sea slugs, a favorite experimental animal among neuroscientists. To his pleasant surprise, CPEB was present in the slug's neurons. Furthermore, CPEB was present precisely where a synaptic marker *should* be, silently skulking in the dendritic branches.

Si and Kandel were intrigued. They now tried to understand CPEB by blocking it. If CPEB was removed, could the neuron make a memory? Could the cell still mark a synapse? Though they hardly believed the data, the answer was clear: without CPEB, the slug's neurons were unable to remember *anything.*

But he still couldn't figure out how CPEB worked. How did this molecule exist outside time? What made it so strong? How did it survive the merciless climate of the brain? Si's first clue arrived when he decoded the protein's amino acid sequence. Most proteins

* Before Dr. Si, the conventional explanation for long-term memory revolved around CREB, a gene activated in neurons during Pavlovian conditioning. But CREB's effects are cell-wide, so it could not explain the formation of memory at specific dendrites.

read like a random list of letters, their structures a healthy mix of different amino acids. CPEB however, looked completely different. One end of the protein had a weird series of amino acid repetitions, as if its DNA had had a stuttering fit (Q stands for the amino acid glutamine):

QQQLQQQQQQQBQLQQQQ

Immediately, Si began looking for other molecules with similar odd repetitions. In the process, he stumbled into one of the most controversial areas of biology. He found what looked like a prion.

Prions were once regarded as the nasty pathogens of a tribe of the worst diseases on earth: mad cow disease, fatal familial insomnia (whose victims lose the ability to sleep, and after three months die of sleep deprivation), and a host of other neurodegenerative diseases. Prions are still guilty of causing these horrific deaths. But biologists are also beginning to realize that prions are everywhere. Prions are roughly defined as a class of proteins that can exist in two functionally distinct states (every other protein has only one natural state). One of these states is active and one is inactive. Furthermore, prions can switch states (turn themselves on and off) without any guidance from above; they change proteomic structure without changing DNA. And once a prion is turned on, it can transmit its new, infectious structure to neighboring cells with no actual transfer of genetic material.

In other words, prions violate most of biology's sacred rules. They are one of those annoying reminders of how much we don't know. Nevertheless, prions in the brain probably hold the key to changing our scientific view of memory. Not only is the CPEB protein sturdy enough to resist the effects of the clock — prions are famous for being virtually indestructible — but it displays an astonishing amount of plasticity. Free from a genetic substrate, CPEB prions are able to change their shapes with relative ease, creating or erasing a memory. Stimulation with serotonin or dopamine, two neurotransmitters that are released by neurons when you think, changes the very structure of CPEB, switching the protein into its active state.

After CPEB is activated, it marks a specific dendritic branch as a memory. In its new conformation, it can recruit the requisite mRNA needed to maintain long-term remembrance. No further stimulation or genetic alteration is required. The protein will patiently wait, quietly loitering in your synapses. One could never eat another madeleine, and Combray would still be there, lost in time. It is only when the cookie is dipped in the tea, when the memory is summoned to the shimmering surface, that CPEB comes alive again. The taste of the cookie triggers a rush of new neurotransmitters to the neurons representing Combray, and, if a certain tipping point is reached, the activated CPEB infects its neighboring dendrites. From this cellular shudder, the memory is born.

But memories, as Proust insisted, don't just stoically endure: they also invariably change. CPEB supports Proust's hypothesis. Every time we conjure up our pasts, the branches of our recollections become malleable again. While the prions that mark our memories are virtually immortal, their dendritic details are always being altered, shuttling between the poles of remembering and forgetting. The past is at once perpetual and ephemeral.

This rough draft of a theory has profound implications for the neuroscience of memory. First of all, it's proof that prions are not some strange biological apocrypha. In reality, prions are an essential ingredient of life and have all sorts of intriguing functions. Swiss scientists, following up on the research of Si and Kandel, have even discovered a link between the prion gene that causes mad cow disease and increased long-term memory. Essentially, the more likely your neurons are to form misfolded prions, the better your memory is. Other experiments have linked a lack of CPEB in the mouse hippocampus to specific deficits in long-term memory. Though the details remain mostly obscure, there seems to be a deep connection between prions and remembrance.

But the CPEB model also requires that we transform our metaphors for memory. No longer can we imagine memory as a perfect mirror of life. As Proust insisted, the remembrance of things past is not necessarily the remembrance of things as they were. Prions reflect this fact, since they have an element of randomness built into their structure. They don't mind fibbing. While CPEB can switch to an active state under a given set of experimental circumstances (like a few puffs of serotonin), Si's experiments also show that the protein can become active for no real reason, since its transformation is largely dictated by the inscrutable laws of protein folding and stoichiometry. Like memory itself, CPEB delights in its contingency.

This indeterminacy is part of CPEB's design. For a protein, prions are uniquely liberated. They are able to ignore everything from the instructions of our DNA to the life cycles of our cells. Though they exist inside us, they are ultimately apart from us, obeying rules of their own making. As Proust said, "The past is hidden . . . in some material object of which we have no inkling."

And though our memory remains inscrutable, the CPEB molecule (if the theory is true) is the synaptic detail that persists outside time. Dr. Si's idea is the first hypothesis that begins to explain how sentimental ideas endure. It is why Combray can exist silently below the surface, just behind the curtain of consciousness. It is also why Marcel remembers Combray on page 58, and not on page 1. It is a molecular theory of explicit memory that *feels* true. Why? Because it embraces our essential randomness, because prions are by definition unpredictable and unstable, because memory obeys nothing but itself. This is what Proust knew: the past is never past. As long as we are alive, our memories remain wonderfully volatile. In their mercurial mirror, we see ourselves.

Chapter 5

Paul Cézanne

The Process of Sight

> How can one learn the truth by thinking? As one learns
> to see a face better if one draws it.
>
> — Ludwig Wittgenstein

IN HER ESSAY "Character in Fiction," Virginia Woolf grandly declared "that on or about December 1910 human nature changed." She was being only a little ironic. The seemingly arbitrary date was an allusion to the first exhibition of postimpressionist paintings.* Paul Cézanne was the star of the show. His revolutionary pictures were mostly of modest things: fruit, skulls, the parched Provençal landscape. These humble subjects only highlighted Cézanne's painterly form, his blatant abandonment of what Roger Fry, in his introduction to the exhibit, called "the cliché of representation." No longer, Fry declared, would "art aim at a pseudo-scientific fidelity to appearance. This is the revolution that Cézanne has inaugurated . . . His paintings aim not at illusion or abstraction, but at reality."

It is not easy to change the definition of reality. At the 1910 exhibition, Cézanne's paintings were denounced in the press as "being

---

* Woolf was also alluding to the publication of Bertrand Russell's *Principia Mathematica*. Russell's epic work of logic provoked a firestorm of debate about whether reality was reducible to its logical fundamentals. For details, see *The Phantom Table*, by Anne Banfield.

of no interest except for the student of pathology and the specialist in abnormality." Cézanne, the critics declared, was literally insane. His art was nothing more than an ugly untruth, a deliberate distortion of nature. The academic style of painting, with its emphasis on accurate details and fine-grained verisimilitude, refused to fade away.

This conservative aesthetic had scientific roots. The psychology of the time continued to see our senses as perfect reflections of the outside world. The eye was like a camera: it collected pixels of light and sent them passively on to the brain. The founder of this psychology was the eminent experimentalist William Wundt, who insisted that every sensation could be broken down into its simpler sense data. Science could peel back the layers of consciousness and reveal the honest stimuli underneath.

Cézanne inverted this view of vision. His paintings were about the subjectivity of sight, the illusion of surfaces. Cézanne invented postimpressionism because the impressionists just weren't strange enough. "What I am trying to translate," Cézanne said, "is more mysterious; it is entwined in the very roots of being." Monet and Renoir and Degas believed that sight was simply the sum of its light. In their pretty paintings they wanted to describe the fleeting photons absorbed by the eye, to describe nature entirely in terms of its illumination. But Cézanne believed that light was only the beginning of seeing. "The eye is not enough," he declared. "One needs to think as well." Cézanne's epiphany was that our impressions require interpretation; to look is to create what you see.

We now know that Cézanne was right. Our vision begins with photons, but this is only the beginning. Whenever we open our eyes, the brain engages in an act of astonishing imagination, as it transforms the residues of light into a world of form and space that we can understand. By probing inside the skull, scientists can see how our sensations are created, how the cells of the visual cortex silently construct sight. Reality is not out there waiting to be witnessed; reality is made by the mind.

Cézanne's art exposes the process of seeing. Although his paintings were criticized for being unnecessarily abstract — even the impressionists ridiculed his technique — they actually show us the world as it first appears to the brain. A Cézanne picture has no boundaries or stark black lines separating one thing from the next. Instead, there are only strokes of paint, and places on the canvas where one color, knotted on the surface, seems to change into another color. This is the start of vision: it is what reality looks like before it has been resolved by the brain. The light has not yet been made into form.

But Cézanne did not stop there. That would have been too easy. Even as his art celebrates its strangeness, it remains loyal to what it represents. As a result, we can always recognize Cézanne's subjects. Because he gives the brain just enough information, viewers are able to decipher his paintings and rescue the picture from the edge of obscurity. (His forms might be fragile, but they are never incoherent.) The layers of brushstrokes, so precise in their ambiguity, become a bowl of peaches, or a granite mountain, or a self-portrait.

This is Cézanne's genius: he forces us to see, in the same static canvas, the beginning and the end of our sight. What starts as an abstract mosaic of color becomes a realistic description. The painting emerges, not from the paint or the light, but from somewhere inside our mind. We have entered into the work of art: its strangeness is our own.

Cézanne never lived to see culture and science catch up with his avant-garde. He was a postimpressionist before impressionism was fully accepted. But for Fry and Woolf, Cézanne's style seemed prophetically modern. In the autumn of 1912, six years after Cézanne died alone in Provence, Fry mounted the second postimpressionist exhibition at the Grafton Gallery. Cézanne's paintings were now seen as the start of a serious movement; his artistic experiments were no longer lonely. The white walls also displayed canvases by Matisse; the New Russians; and Virginia's sister, Vanessa Bell. Abstraction had become the new realism.

The Invention of the Photograph

The story of abstract painting begins with the photograph, which literally means "light writing." That's what a photograph is: an image written in frozen light. Ever since the Renaissance, artists have used camera obscuras ("dark rooms") to condense the three planes of reality into two dimensions. Leonardo da Vinci described the instrument in his notebooks as a metaphor for the eye. Giovanni Battista Della Porta, in his 1558 treatise *Magia Naturalis (Natural Magic)*, advocated the camera as a tool for struggling painters.

But it wasn't until the nineteenth century, with the discovery of photosensitive chemicals, that painting lost its monopoly on representation. Verisimilitude was now a technology. Louis Daguerre, a commercial painter, was the inventor of the photographic plate. By treating silver-coated copper sheets with iodine, Daguerre created a flat surface sensitive to light. He then exposed these plates in a primitive camera (a black box with a hole) and developed the images with the warm poison of mercury vapor. The pixels emerged like accurate ghosts. By immersing the plate in a salt solution, Daguerre made the ghosts permanent. Light had been captured.

Painters, still in the business of copying reality, saw the new technology as a dire threat. How could the human hand compete with the photon? J.M.W. Turner is said to have remarked after seeing a daguerreotype that he was glad he'd already had his day, since the era of painting was now over. But not all artists believed in the inevitable triumph of the camera. The symbolist poet Charles Baudelaire, a natural skeptic of science, reviewed a photographic exhibition in 1859 by proclaiming the limits of the new medium. Its accuracy, he said, is deceptive, nothing more than phony simulacra of what was really out there. The photographer was even — and Baudelaire only used this insult in matters of grave import — a *materialist*. In Baudelaire's romantic view, the true duty of photography was "to be the servant of the sciences and arts, but the very humble servant, like printing or shorthand, which have neither created nor supplemented literature. . . . If it [photography] is allowed to encroach upon the domain of the imaginary, upon anything whose value depends solely upon the addition of something of a man's soul, then it will be so much the worse for us." Baudelaire wanted the modern artist to describe everything that the photograph ignored: "the transient, the fleeting, the contingent."

Inspired by Baudelaire's writings and the provocative realism of Edouard Manet, a motley group of young French painters decided to rebel. The camera, they believed, was a liar. Its precision was false. Why? Because reality did not consist of static images. Because the camera stops time, which cannot be stopped; because it renders everything in focus, when everything is never in focus. Because the eye is not a lens, and the brain is not a machine.

These rebels called themselves the impressionists. Like the film in a camera, their idiom was light. But the impressionists realized that light was both a dot and a blur. If the camera captured the dot, the impressionists represented the blur. They wanted to capture *time* in their paintings, showing how a bale of hay changes in the afternoon shadows, or how the smoke of a train leaving Gare Saint-Lazare slowly fades into thin air. As Baudelaire insisted, they painted what the camera left out.

Look, for example, at an early Monet, *Impression: Soleil Levant (Impression: Sunrise)*. Monet painted this hazy scene of the Le Havre harbor in the spring of 1872. An orange sun hangs in a gray sky; a lonely fisherman sails in a sea made of undulating brushstrokes. There is little here for the eye to see. Monet is not interested in the ships, or in their billowing sails, or in the glassy water. He ignores all the static things a photograph would detect. Instead, Monet is interested in the moment, in its transience, in his impression of its transience. His mood is mixed in with the paint, his subjectivity muddled with the facts of his sensations. This, he seems to be saying, is a scene no photograph could catch.

With time, the impressionists grew more radical. This was partly due to eye troubles: Monet became blind (but didn't stop painting the bridges of Giverny). Vincent van Gogh, drinker of kerosene, turpentine, and absinthe, probably thought the coronas he painted around stars and streetlamps were real. Edgar Degas became severely myopic, which led him to do more and more sculpture ("I must learn a blind man's trade now," Degas said). Auguste Renoir, poisoned by his pastel paints, became a rheumatic cripple.

But whether their abstraction was motivated by physiology or philosophy, it became increasingly clear that the impressionists had broken with the staid traditions of academic realism.* They didn't paint religious heroes or epic battles or portraits of the royal family. Instead, they continued to paint what they had always painted: the Sunday picnics of the bourgeoisie and women in bathtubs and purple lilies floating on light-dappled water. When the critics ridiculed their work as frivolous and false, they just shrugged. After all, the impressionist art was a celebration of technique; wherever there was light, they could create paintings.

This is why the impressionists feel modern, while Delacroix and Ingres and Bouguereau do not: they realized the painter did not simply have a subject that he or she was duty bound to represent. The painter was an artist, and artists had ideas that they were compelled to *express*. In their unsellable canvases — the Louvre wouldn't even accept them as gifts — the impressionists invented the idea of painterly abstraction. Color became symbolic. Blurriness was chic. The gaze was out, the glance was in.

But the thing about art movements is that they are always moving. By freeing the artist from the strict limits of verisimilitude, impressionism was destined go places those water lilies could never have imagined. And if Monet and Degas, prodded by the camera, led the way into impressionism, Paul Cézanne led the way into its aftermath. As he immodestly declared at the beginning of his ca-

---

* The stylistic innovations of the impressionists also depended on developments in paint technology. For example, the cobalt violet that Monet often used to paint the ocean and sky had been invented just a few years earlier by industrial chemists. Monet quickly realized that this new pigment had enormous potential for describing the effects of light. "I have finally discovered the color of the atmosphere," Monet declared. "It is violet."

reer, "I want to make of impressionism something solid and lasting, like the art in the museums."

Cézanne often spent hours contemplating a brushstroke. Out in the open air, he would stare at his subject until it melted under his gaze, until the forms of the world had decayed into a formless mess. By making his vision disintegrate, Cézanne was trying to return to the start of sight, to become nothing but "a sensitive recording plate." The slowness of this method forced Cézanne to focus on simple things, like a few red apples set on a trapezoid of table, or a single mountain seen from afar. But he knew that the subject itself was irrelevant. Stare hard enough, his paintings implore, and the laws of the known universe will emerge from just about anything. "With an apple," Cézanne once said, "I will astonish Paris."

The founder of postimpressionism learned to paint from a quintessential impressionist: Camille Pissarro. The two made an incongruous pair. Pissarro was a French Creole Jew from the West Indies, while Cézanne was a coarse — some said crude — Provençal. Their friendship was founded upon their shared sense of isolation. Both were exiles from the academic style of the time, which had made Ingres into a god and talent synonymous with fine-grained resolution. Pissarro and Cézanne had neither the temperament nor the patience for such art. Pissarro was a friendly anarchist and recommended burning down the Louvre, while Cézanne — speaking of his early painting instructors — declared, "Teachers are all castrated bastards and assholes. They have no guts."

Alone together, Pissarro and Cézanne saturated themselves in their own style. Cézanne would methodically copy Pissarro's paintings in order to understand his impressionist technique. "The eye must absorb everything," Pissarro instructed him. "Do not follow rules and principles, but paint what you see and feel. And do not be shy of nature." Cézanne listened to Pissarro. Before long, the burnt umbers and mahoganies of Cézanne's early paintings (he loved Courbet) had become the layers of pastel typical of impression-

Green Apples, *by Paul Cézanne, 1873*

ism. An early work, *Rocks at L'Estaque,* depicting Cézanne's favorite Provençal landscape, clearly demonstrates Pissarro's influence. Staccato brushstrokes predominate; the colors are basic but exist in myriad tones. Depth and structure, even the time of day: all are defined by minute differentiations in the tint of paint. But *Rocks at L'Estaque,* for all its impressionist splendor, also shows Cézanne inventing his own avant-garde.

This is because Cézanne had stopped worshipping light. He found the impressionist project — the description of light's dance upon the eye — too insubstantial. ("Monet is only an eye," Cézanne once said, with more than a little condescension.) In the *Rocks at L'Estaque,* the sea in the distance does not sparkle as Pissarro would have had it sparkle. The granite does not glint in the sunshine, and nothing casts a shadow. Cézanne was not interested, as the impressionists were, in reducing everything to surfaces of light. He had stopped arguing with the camera. Instead, in his postimpressionist paintings he wanted to reveal how the moment is *more* than its light. If the impressionists reflected the eye, Cézanne's art was a mirror held up to the mind.*

What did he see in the mirror? Cézanne discovered that visual forms — the apple in a still life or the mountain in a landscape — are mental inventions that we unconsciously impose onto our sensations. "I tried to copy nature," Cézanne confessed, "but I couldn't. I searched, turned, looked at it from every direction, but in vain." No matter how hard he tried, Cézanne couldn't escape the sly interpretations of his brain. In his abstract paintings, Cézanne wanted to reveal this psychological process, to make us aware of the particular way the mind creates reality. His art shows us what we cannot see, which is how we see.

The Limits of Light

Understanding how sight starts, how the eyeball transforms light into an electrical code, is one of the most satisfying discoveries of modern neuroscience. No other sense has been so dissected. We now know that vision begins with an atomic disturbance. Particles of light alter the delicate molecular structure of the receptors in the retina. This cellular shudder triggers a chain reaction that ends with a flash of voltage. The photon's energy has become information.†

But that code of light, as Cézanne knew, is just the start of seeing. If sight were simply the retina's photoreceptors, then Cézanne's canvases would be nothing but masses of indistinct color. His Provençal landscapes would consist of meaningless alternations of olive and ocher, and his still lifes would be all paint and no fruit. Our

---

* Emile Bernard, a French painter, was one of the few people to actually observe Cézanne's painting technique. As he watched Cézanne construct his canvases, Bernard was struck by how Cézanne had broken entirely with the rules of impressionism: "His method was not at all like that . . . Cézanne only interpreted what he saw, he did not try to copy it. His vision was centered much more in his brain, than in his eye" (Doran, p. 60).

† The electrical message of photoreceptors is actually the absence of an electrical message, as the photons cause the sodium ion channels inside our photoreceptors to close, which causes the cell to become hyperpolarized. Eyes speak with silence.

world would be formless. Instead, in our evolved system, the eye-ball's map of light is transformed again and again until, millisec-onds later, the canvas's description enters our consciousness. Amid the swirl of color, we see the apple.

What happens during this blink of unconscious activity? The first scientific glimpse into how the brain processes the eye's data arrived in the late 1950s, in an astonishing set of experiments by Da-vid Hubel and Torsten Wiesel. At the time, neuroscience had no idea what kind of visual stimuli the cortex responded to. Light ex-cites the retina, but what kind of visual information excites the mind? The experiments attempting to answer this question were brutally simple: points of light were flashed onto an animal's retina (a poor cat was usually used) while a galvanic needle recorded cel-lular electricity from a brain region called the V1, which is the first stage of our visual cortex. If some voltage was detected, then the cell was seeing something. Before Hubel and Weisel, scientists assumed that the eye was like a camera, and that the brain's visual field was composed of dots of light, neatly arranged in time and space. Just as a photograph was made up of a quilt of pixels, so must the eye create a two-dimensional representation of reflected light that it seamlessly transmitted to the brain. Yet when scientists tried find-ing this camera inside the skull, all they found was silence, the elec-trical stupor of uninterested cells.

This was a frustrating paradox. The animal clearly could see, and yet its cells, when isolated with a beam of light, were quiet. It was as if the animal's vision was emerging from a blank canvas. Hubel and Weisel bravely ventured into this mystery. At first, their results only confirmed the impossibility of activating cortical neurons with in-dividual pricks of light. But then, by complete accident, they discov-ered an excited cell, a neuron interested in the slice of world it had seen.

What was this cell responding to? Hubel and Weisel had no idea. The neuron became active at the exact moment it was supposed to be silent, when they were in between experiments. There was no light to excite it. Only after retracing their exact steps did Hubel and

Wiesel figure out what had happened. As they had inserted a glass slide into the light projector, they had inadvertently cast "a faint but sharp shadow" onto the cat's retina. It was just a fleeting glint of brightness — a straight line pointed in a single direction — but it was exactly what the cell wanted.

Hubel and Wiesel were stunned by their discovery. They had glimpsed the raw material of vision, and it was completely abstract. Our brain cells were strange things, fascinated not by dots of light but by angles of lines.* These neurons preferred contrast over brightness, edges over curves. In their seminal 1959 paper "Receptive Fields of Single Neurons in the Cat's Striate Cortex," Hubel and Wiesel became the first scientists to describe reality as it appears to the early layers of the visual cortex. This is what the world looks like before it has been seen, when the mind is still creating the sense of sight.

Cézanne's paintings echo this secret geometry of lines sensed by the visual cortex. It's as if he broke the brain apart and saw how seeing occurs. Look, for example, at *The Rocks Above the Caves at Château Noir* (1904–1906). Cézanne has chosen a typically simple subject, just a few boulders surrounded by some scraggly trees. Windows of blue sky break through the foliage. But Cézanne's painting is not about the sky or the rocks or the trees. He has broken each of these elements into their sensory parts, deconstructing the scene in order to show us how the mind reconstructs it.

At the literal level of paint, Cézanne represented the landscape as nothing but a quilt of brushstrokes, each one a separate line of color. He abandoned the pointillism of Seurat and Signac, in which everything is dissected into discrete points of light. Instead, Cézanne pursued a much more startling path, creating the entire picture out of patches and strokes, *les tâches* and *les touches*. His im-

---

* The early parts of our visual cortex are stimulated by visual inputs that look very similar to a Piet Mondrian painting. Mondrian, a painter extremely influenced by Cézanne, spent his life searching for what he called "the constant truths concerning forms." He eventually settled on the straight line as the essence of his art. He was right, at least from the perspective of the V1.

pasto paint calls attention to itself, forcing us to see the canvas as a constructive process and not a fixed image. As the art historian Meyer Schapiro noted, in a Cézanne painting "it is as if there is no independent, closed, pre-existing object, given to the painter's eye for representation, but only a multiplicity of successively probed sensations." Instead of giving us a scene of fully realized forms, Cézanne supplies us with layers of suggestive edges, out of which forms slowly unfurl. Our vision is made of lines, and Cézanne has made the lines distressingly visible.

This is the abstract reality represented by the neurons of the V1. As the surface of Cézanne's painting testifies, our most elemental level of sensation is replete with contradiction and confusion. The cells of the visual cortex, flooded by rumors of light, see lines extending in every possible direction. Angles intersect, brushstrokes disagree, and surfaces are hopelessly blurred together. The world is still formless, nothing but a collage of chromatic blocks. But this ambiguity is an essential part of the seeing process, as it leaves space for our subjective interpretations. Our human brain is designed so that reality cannot resolve itself. Before we can make sense of Cézanne's abstract landscape, the mind must intervene.

So far, the story of sight has been about what we actually sense: the light and lines detected by the retina and early stages of the visual cortex. These are our feed-forward projections. They represent the external world of reflected photons. And while seeing begins with these impressions, it quickly moves beyond their vague suggestions. After all, the practical human brain is not interested in a camera-like truth; it just wants the scene to make sense. From the earliest levels of visual processing in the brain up to the final polished image, coherence and contrast are stressed, often at the expense of accuracy.

Neuroscientists now know that what we end up seeing is highly influenced by something called top-down processing, a term that describes the way cortical brain layers project down and influence (corrupt, some might say) our actual sensations. After the inputs of the eye enter the brain, they are immediately sent along two separate pathways, one of which is fast and one of which is slow. The fast pathway quickly transmits a coarse and blurry picture to our prefrontal cortex, a brain region involved in conscious thought. Meanwhile, the slow pathway takes a meandering route through the visual cortex, which begins meticulously analyzing and refining the lines of light. The slow image arrives in the prefrontal cortex about fifty milliseconds after the fast image.

Why does the mind see everything twice? Because our visual cortex needs help. After the prefrontal cortex receives its imprecise picture, the "top" of the brain quickly decides what the "bottom" has seen and begins doctoring the sensory data. Form is imposed onto the formless rubble of the V1; the outside world is forced to conform to our expectations. If these interpretations are removed, our reality becomes unrecognizable. The light just isn't enough.

The neurologist Oliver Sacks once had a patient, Dr. P, who inhabited a world that looked like a Cézanne canvas. Due to a cortical lesion, Dr. P's eyes received virtually no input from his brain. He saw the world solely in its unprocessed form, as labyrinths of light and masses of color. In other words, he saw reality as it actually was. Unfortunately, this meant that his sensations were completely surreal. To explore his patient's illness, Sacks asked Dr. P to describe some photographs in *National Geographic:*

"His [Dr. P's] responses here were very curious. His eyes would dart from one thing to another, picking up tiny features, individual features, as they had done with my face. A striking brightness, a colour, a shape would arrest his attention and elicit comment — but in no case did he get the scene-as-a-whole. He had no sense whatever of a landscape or scene."

Dr. P's problem lay in what happened to the light once it traveled beyond his retina. His eyes were fine; they were absorbing photons perfectly. It was only because his brain couldn't interpret his sensations that he saw the world as such a hopeless commotion of fragments. A photograph seemed abstract. He couldn't recognize his own reflection. Sacks describes what happened when Dr. P got up to leave his office: "He [Dr. P] then started to look round for his hat. He reached out his hand, and took hold of his wife's head, tried to lift it off, to put it on. He had apparently mistaken his wife for a hat! His wife looked as if she was used to such things."

Sacks' tragicomic vignette exposes an essential element of the seeing process. One of the functions of top-down processing is object recognition. The instructions of the prefrontal cortex allow us to assimilate the different elements of an object — all those lines and edges seen by the V1 — into a unified *concept* of the object. This was what Dr. P couldn't do. His impressions of light never congealed into a thing. As a result, before Dr. P could "see" a glove, or his left foot, or his wife, he had to painstakingly decipher his own sensations. Every form needed to be methodically analyzed, as if it were being seen for the first time. For example, when Dr. P was given a rose, he described his conscious thought process to Sacks: "It looks about six inches in length. A convoluted red form with a linear green attachment." But these accurate details never triggered the idea of a rose. Dr. P had to smell the flower before he could identify its form. As Sacks put it, "Dr. P saw nothing as familiar. Visually, he was lost in a world of lifeless abstractions."

To look at a Cézanne painting is to become acutely aware of what Dr. P is missing. Staring at his postimpressionist art, we feel our top-down process at work. It is because Cézanne knew that the impression was not enough — that the mind must *complete* the impression — that he created a style both more abstract and more truthful than the impressionists. And even though his postimpressionist style was seen as needlessly radical — Manet referred to him as "the bricklayer who paints with a trowel" — it wasn't. Cézanne abstracted on nature because he realized that *everything* we see is an abstraction. Before we can make sense of our sensations, we have to impress our illusions upon them.

In his art, Cézanne made this mental process self-evident. While he deconstructed his paintings until they were on the verge of unraveling, his paintings don't unravel, and that is their secret. Instead, they tremble on the edge of existence, full of fractures and fissures that have to be figured out. Such an exquisite balancing act isn't easy. Until Cézanne sold a canvas — and he rarely sold anything — he continued to edit his brushstrokes, trying to edge closer to the delicate reality he wanted to describe. His work would become thick with paint, with layer after layer of carefully applied color. Then the paint would crack, broken by its own mass.

Why was painting such a struggle for Cézanne? Because he knew that a single false brushstroke could ruin his canvas. Unlike the impressionists, who wanted their paintings to reflect the casual atmospherics of being *en plein air*, Cézanne's art was adamantly difficult. In his clenched canvases, he wanted to give the brain just enough to decipher, and not a brushstroke more. If his representations were too accurate or too abstract, everything fell apart. The mind would not be forced to enter the work of art. His lines would have no meaning.

Cézanne and Zola

The year was 1858. Cézanne was eighteen. His best friend, Emile Zola, had just left for Paris, leaving him behind in Aix-en-Provence. Zola had already decided to become a writer, but Cézanne, following the demands of his authoritarian father, was busy failing out of law school. Zola was furious with Cézanne. "Is painting only a whim for you?" he angrily asked. "Is it only a pastime, a subject of conversation? If this is the case, then I understand your conduct: you are right not to make trouble with your family. But if painting is your vocation, then you are an enigma to me, a sphinx, someone impossible and obscure." The very next summer, Cézanne fled to Paris. He had decided to become an artist.

Life in the city was difficult. Cézanne was lonely and impoverished. Being a bohemian was overrated. During the day, he sneaked into the Louvre, where he patiently copied works by Titian and Rubens. At night, everyone crowded into the local bar and drunkenly argued about politics and art.

Cézanne felt like a failure. His first experiments in abstraction were dismissed as accidental mistakes, the feeble work of a talentless realist. He carted his paintings around the city in a wheelbarrow, but no gallery would accept them. Cézanne's only consolation lay in the culture at large: the stuffy Parisian art world was finally beginning to change. Baudelaire had begun assailing Ingres. Manet was studying Velazquez. The gritty paintings of Gustave Courbet — his mantra was "Let us be true, even if we are ugly" — were slowly gaining respect.

By 1863, all of this new "ugliness" could no longer be suppressed. There was simply too much of it. As a result, Emperor Napoleon III decided to exhibit the paintings rejected by the Academy of Fine Arts for their annual art show. It was in this gallery — the Salon of the Refused — that Cézanne would first glimpse Manet's *Le Déjeuner sur l'Herbe (The Luncheon on the Grass)*, a scandalous picture of a naked woman in a park who doesn't seem to know she's naked. Cézanne was mesmerized. He began a series of paintings in which he re-imagined Manet's pornographic picnic. Unlike Manet, who painted the woman with a sense of ironic detachment, Cézanne inserted himself into the middle of the artwork. His scraggly beard and bald head give him away. The same critics that had been disdainful of Manet were now cruel to Cézanne. "The public sneers at this art," wrote one reviewer. "Mr. Cézanne gives the impression of being a sort of madman who paints in *delirium tremens*. He must stop painting."

Twenty years later, everything had changed. The emperor was gone, defeated in battle during the Franco-Prussian War. Claude Monet, who fled Paris in order to avoid serving in the army, had glimpsed the prophetically abstract watercolors of J.M.W. Turner while in

London. He returned to France newly inspired. By 1885, Monet's impressionism was a genuine avant-garde. The painters of hazy light now had their own salons.

The intervening years had also been kind to Zola. His Rougon-Macquart novels turned him into a literary celebrity, confidently controversial. He was the proud founder of naturalism, a new school of literature that aspired to write "the scientific novel." The novelist, Zola declared, must literally become a scientist, "employing the experimental method in their study of man."

Flush with his success, Zola decided to write a book about a painter. He called it *L'Oeuvre (The Masterpiece)* because he said he could think of nothing better. As required by his method, Zola based his fiction on a story stolen straight from real life. The life he stole this time was the life of his best friend. After the novel was published, in the spring of 1886, Cézanne and Zola never spoke to each other again.

The protagonist of *The Masterpiece* is Claude Lantier. Like Cézanne, he is a bearded and balding Provençal, a painter whose paintings are too strange to display. Zola even got the afflictions right: both Claude and Paul suffer from incurable eye diseases, are ridiculed by their fathers, and have to trade their paintings for food at the local grocery. While Claude is the stereotypical struggling artist, his best friend, the thinly veiled writer Pierre Sandoz, has achieved great literary acclaim, writing a series of twenty novels documenting "the truth of humanity in miniature."

But the real insult came when Zola described Claude's art. His abstract paintings, Zola wrote, were nothing but "wild mental activity . . . the terrible drama of a mind devouring itself." Sandoz's novels, on the other hand, "describe man as he really is." They are a "new literature for the coming century of science."

It was clear that Zola had betrayed his impressionist friends. Monet, Pissarro, and the symbolist poet Stéphane Mallarmé held meetings to denounce the book. "Our enemies," Monet wrote to Zola, "will make use of your book to cudgel us senseless." But Zola didn't care. He had turned against abstraction. If Cézanne's paintings made our subjectivity their subject, Zola's novels were determined to turn man into just another object. The artist, Zola said, must "disappear, and simply show what he has seen. The tender intervention of the writer weakens a novel, and introduces a strange element into the facts which destroys their scientific value."

Zola's style didn't last long. His self-proclaimed scientific novels, with their naïve faith in heredity and biological determinism, aged ungracefully. His work was not the "immortal encyclopedia of human truths" he had expected it to become. As Oscar Wilde declared, "Zola sits down to give us a picture of the Second Empire. Who cares for the Second Empire now? It is out of date. Life goes faster than Realism." Even worse, the avant-garde that Zola betrayed in *The Masterpiece* was now ascendant. His movement was being usurped by postimpressionism. By 1900, Zola was forced to admit that he had misjudged Cézanne's abstract art. "I have a better understanding of his painting," Zola confessed, "which eluded me for a long time because I thought it was exaggerated, but actually it is unbelievably sincere and truthful."

In the end, it was not *The Masterpiece* that drove Cézanne and Zola apart. Zola never apologized, but it was just as well: no apology could heal the rift in their philosophies. They were two childhood friends who had come to conflicting conclusions about the nature of reality. If Zola tried to escape himself in his art — fleeing instead into the cold realm of scientific fact — Cézanne sought reality by venturing *into* himself. He knew that the mind makes the world, just as a painter makes a painting.

With that startling revelation, Cézanne invented modernist art. His canvases were deliberately new; he broke the laws of painting in order to reveal the laws of seeing. If he left some details out, it was only to show us what we put in. Within a few decades, of course, Paris would be filled with a new generation of modern painters who liked to break the law even more. The cubists, led by Pablo Picasso and Georges Braque, would take Cézanne's technique to its incongruous conclusion. (Picasso once declared that Cézanne and

Buffalo Bill were his two greatest influences.) And even though the cubists liked to joke about anticipating the weird facts of quantum physics, no other painter got the human mind like Cézanne. His abstractions reveal our anatomy.

The Blank Canvas

As Cézanne aged, his paintings became filled by more and more naked canvas, what he eloquently called *nonfinito*. No one had ever done this before. The painting was clearly incomplete. How could it be art? But Cézanne was unfazed by his critics. He knew that his paintings were only literally blank.* Their incompleteness was really a metaphor for the process of sight. In these unfinished canvases, Cézanne was trying to figure out what the brain would finish for him. As a result, his ambiguities are exceedingly deliberate, his vagueness predicated on precision. If Cézanne wanted us to fill in his empty spaces, then he had to get his emptiness exactly right.

For example, look at Cézanne's watercolors of Mont Sainte-Victoire. In his final years, Cézanne walked every morning to the crest of Les Lauves, where an expansive view of the Provençal plains opened up before him. He would paint in the shade of a linden tree. From there, Cézanne said, he could see the land's hidden patterns, the way the river and vineyards were arranged in overlapping planes. In the background was always the mountain, that jagged isosceles of rock that seemed to connect the dry land with the infinite sky.

Cézanne, of course, was not interested in literal portraits of the landscape. In his descriptions of the valley, Cézanne wanted to paint only the essential elements, the necessary skeleton of form. And so he summarized the river in a single swerve of blue. The groves of chestnut trees became little more than dabs of dull green, interrupted occasionally by a subtle stroke of umber. And then

* As Gertrude Stein said of one of these Cézanne landscapes, "Finished or unfinished, it always was what looked like the very essence of an oil painting, because everything was there."

Mount Sainte-Victoire Seen from Lauves, *by Paul Cézanne, 1904–1905*

there was the mountain. Cézanne often condensed the foreboding mass of Mont Sainte-Victoire into a single line of dilute paint, dragged across the empty sky. This thin gray line — the shadowy silhouette of the mountain — is completely surrounded by negative space. It is a fragile scratch against the sprawling void.

And yet the mountain does not disappear. It is *there*, an implacable and adamant presence. The mind easily invents the form that Cézanne's paint barely insinuates. Although the mountain is almost literally invisible — Cézanne has only implied its presence — its looming gravity anchors the painting. We don't know where the painting ends and we begin.

Cézanne's embrace of the blank canvas — his decision to let the emptiness show through — was his most radical invention. Unlike the academic style, which worshipped clarity and decorative detail above all, the subject of Cézanne's postimpressionist paintings was their own ambiguity. With their careful confusion of things and nothing, Cézanne's *nonfinito* paintings question the very essence of form. His incomplete landscapes were proof that even when there was no sensation — the canvas was empty — we could still see. The mountain was still there.

When Cézanne began his studies in the blank canvas, science had no way of explaining why the paintings appeared less vacant than they actually were. The very existence of Cézanne's *nonfinito* style, the fact that the brain could find meaning in nothing, seemed to disprove any theory of mind that reduced our vision to pixels of light.

The Gestalt psychologists of the early twentieth century were the first scientists to confront the illusions of form that Cézanne so eloquently manipulated. *Gestalt* literally means "form," and that's what interested Gestalt psychologists. Founded by Carl Stumpf, Kurt Koffka, Wolfgang Köhler, and Max Wertheimer in the beginning of the twentieth century, the German Gestalt movement began as a rejection of the reductionist psychology of its time, which was still enthralled with the theories of Wilhelm Wundt and his fellow psychophysicists. Wundt had argued that visual perception is ultimately reducible to its elemental sensations. The mind, like a mirror, reflected light.

But the mind is *not* a mirror. The Gestaltists set out to prove that the process of seeing alters the world we observe. Like Immanuel Kant, their philosophical precursor, they argued that much of what was thought of as being *out there* — in our sensations of the outside world — actually came from *in here,* from inside the mind. ("The imagination," Kant wrote, "is a necessary ingredient of perception itself.") As evidence for their theories of perception, the Gestaltists used optical illusions. These ranged from the illusion of apparent motion in a movie (the film is really a set of static photographs flipped twenty-four times a second) to drawings that seem to oscillate between two different forms (the classic example is the vase that can also be seen as two faces in silhouette). According to the Gestaltists, these everyday illusions were proof that everything we saw was an illusion. Form is dictated from the top down. Unlike the

Wundtians, who began with our sensory fragments, the Gestaltists began with reality as we actually experienced it.

Modern neuroscientific studies of the visual cortex have confirmed the intuitions of Cézanne and the Gestaltists: visual experience transcends visual sensations. Cézanne's mountain arose from the empty canvas because the brain, in a brazen attempt to make sense of the painting, filled in its details. This is a necessary instinct. If the mind didn't impose itself on the eye, then our vision would be full of voids. For example, because there are no light-sensitive cones where the optic nerve connects to the retina, we each have a literal blind spot in the center of the visual field. But we are blind to our own blind spot: our brain unfailingly registers a seamless world.

This ability to make sense of our incomplete senses is a result of human cortical anatomy. The visual cortex is divided into distinct areas, neatly numbered 1 through 5. If you trace the echoes of light from the V1, the neural area where information from the retina first appears as a collection of lines, to the V5, you can watch the visual scene acquire its unconscious creativity. Reality is continually refined, until the original sensation — that incomplete canvas — is swallowed by our subjectivity.

The first area in the visual cortex where neurons respond to both illusory and actual imagery is the V2. It is here that the top part of the mind begins altering the lower levels of sight. As a result, we begin to see a mountain where there is only a thin black line. From this point on, we can't separate our own mental inventions from what really exists. The exact same neurons respond when we actually see a mountain and when we just imagine a mountain. There is no such thing as immaculate perception.

After being quickly processed by the other areas of the visual cortex — color and motion are now integrated into the picture — the data flows into the medial temporal lobe (also known as V5), the region in the brain that gives rise to conscious perceptions. In this area near the back of the head, small subsets of cells first respond to complex stimuli, such as a Cézanne painting of a mountain, or a real mountain. When these specific neurons light up, all the visual processing is finally finished. The sensation is now ready for consciousness.

And because neurons in the temporal cortex are very specific in their representations, tiny lesions in this brain region can erase entire categories of form. This syndrome is called visual-object agnosia. Some victims of this syndrome might fail to perceive apples, or faces, or postimpressionist paintings. Even if the victim maintains an awareness of the object's various elements, he or she is unable to bind those fragments into a coherent representation. The point is that our world of form only exists at this late stage of neural processing, in cranial folds far removed from the honest light of the outside world.

Furthermore, the nerves that feed into consciousness are themselves modulated by consciousness. Once the prefrontal cortex thinks it has seen a mountain, it starts adjusting its own inputs, imagining a form in the blank canvas. (To paraphrase Paul Simon, "A man sees what he wants to see, and disregards the rest.") In fact, in the lateral geniculate nucleus (LGN), the thick nerve that connects the eyeball to the brain, ten times more fibers project from the cortex to the eye than from the eye to the cortex. We make our eyes lie. As William James wrote in *Pragmatism:* "A sensation is rather like a client who has given his case to a lawyer and then has passively to listen in the courtroom to whatever account of his affairs the lawyer finds it most expedient to give."

What is the moral of all these anatomy lessons? The mind is not a camera. As Cézanne understood, seeing is imagining. The problem is that there is no way to quantify what we *think* we see. Each of us is locked inside our own peculiar vision. If we removed our self-consciousness from the world, if we saw with the impersonal honesty of our eyeballs, then we would see nothing but lonely points of light, glittering in a formless space. There would be no mountain. The canvas would simply be empty.

The postimpressionist movement begun by Cézanne was the first style to make our dishonest subjectivity its subject. His paintings are criticisms of paintings: they call attention to their own un-reality. A Cézanne painting admits that the landscape is made of negative space, and that the bowl of fruit is a collection of brushstrokes. Everything has been bent to fit the canvas. Three dimensions have been flattened into two, light has been exchanged for paint, the whole scene has been knowingly composed. Art, Cézanne reminds us, is surrounded by artifice.

The shocking fact is that sight is like art. What we see is not real. It has been bent to fit our canvas, which is the brain. When we open our eyes, we enter into an illusory world, a scene broken apart by the retina and re-created by the cortex. Just as a painter interprets a picture, we interpret our sensations. But no matter how precise our neuronal maps become, they will never solve the question of what we actually see, for sight is a private phenomenon. The visual experience transcends the pixels of the retina and the fragmentary lines of the visual cortex.

It is art, and not science, that is the means by which we express what we see on the inside. The painting, in this respect, is closest to reality. It is what gets us nearest to experience. When we stare at Cézanne's apples, we are inside his head. By trying to represent his own mental representations, Cézanne showed art how to transcend the myth of realism. As Rainer Maria Rilke wrote, "Cézanne made the fruit so real that it ceased to be edible altogether, that's how thinglike and real they became, how simply indestructible in their stubborn thereness." The apples have become what they have always been: a painting created by the mind, a vision so abstract it seems real.

Chapter 6

Igor Stravinsky

The Source of Music

> Only those who will risk going too far can possibly find
> out how far one can go.
>
> — T. S. Eliot

THE BALLET BEGAN at eight. It had been a hot and humid summer day in Paris, and the night remained uncomfortably sultry. The inside of the theater was stifling. As the houselights dimmed, the audience, a little tipsy from their pretheater drinks, put down the programs and stopped murmuring. The men removed their top hats and blotted their foreheads. The ladies unfurled their boas. The curtain slowly rose.

Igor Stravinsky, sweating in his tuxedo in the fourth row, was getting nervous. The ballet *The Rite of Spring*, for which he had written the music, was about to receive its public premiere. An ambitious young composer, Stravinsky was eager to advertise his genius to the cosmopolitan crowd. He wanted his new work of music to make him famous, to be so shockingly new that it could not be forgotten. Modern times demanded modern sounds, and Stravinsky wanted to be the most modern composer around.

The first dance that night was not *The Rite*. Sergei Diaghilev, the impresario of the Ballets Russes and the man who had commissioned Stravinsky's composition, chose to start the evening with a classic, *Les Sylphides*. With piano music by Chopin and choreography by the always graceful Michel Fokine, this crowd-pleasing polonaise represented everything Stravinsky rebelled against. Fokine had been inspired by Chopin's dreamy harmonies and turned the ballet into a reverie of romanticism, a work of pure poetic abstraction. Its only plot was its beauty.

There was no intermission following the piece. After the applause faded, the auditorium filled with a pregnant pause. A few more percussionists crammed into the pit. The string players dutifully retuned their instruments. When they were finished, the conductor, Pierre Monteux, moved his baton to the ready position. He pointed to the bassoon player. *The Rite* had begun.

At first, *The Rite* is seductively easy. The tremulous bassoon, playing in its highest register (it sounds like a broken clarinet), echoes an old Lithuanian folk melody. To the innocent ear, this lilting tune sounds like a promise of warmth. Winter is over. We can hear the dead ground giving way to an arpeggio of green buds.

But spring, as T. S. Eliot pointed out, is also the cruelest time. No sooner do lilacs emerge than the sweeping dissonance of Stravinsky's orchestral work begins, like "the immense sensation that all things experience at the moment when Nature renews its forms." In one of music's most brutal transitions, Stravinsky opens the second section of his work with a monstrous migraine of sound. Though the music has just started, Stravinsky is already relishing the total rejection of our expectations. Stravinsky called this section "The Augurs of Spring."

The "Augurs" don't augur well. Within seconds, the bassoon's flowery folk tunes are paved over by an epileptic rhythm, the horns colliding asymmetrically against the ostinato. All of spring's creations are suddenly hollering for attention. The tension builds and builds and builds, but there is no vent. The irregular momentum is merciless, like the soundtrack to an apocalypse, the beat building to a fatal fortissimo.

This was when the audience at the premiere began to scream. *The Rite* had started a riot.

Once the screaming began, there was no stopping it. After being pummeled by the "Augurs" chord, the bourgeoisie began brawling in the aisles. Old ladies attacked young aesthetes. Insults were hurled at ballerinas. The riot got so loud that Monteux could no longer hear what he was conducting. The orchestra disintegrated into a cacophony of confused instruments. Musical dissonance was usurped by real dissonance. The melee incensed Stravinsky. His art was being destroyed by an idiotic public. His face etched with anger, Stravinsky fled from his seat and ran backstage.

In the wings was Diaghilev, frantically switching the houselights on and off, on and off. The strobe effect only added to the madness. Vaslav Nijinsky, the ballet's choreographer, was just off the stage, standing on a chair and shouting out the beat to the dancers. They couldn't hear him, but it didn't matter. After all, this dance was about the absence of order. Like the music, Nijinsky's choreography was a self-conscious rejection of his art. The refined, three-dimensional shapes of the academic ballet, the discrete positioning of the arms and legs, the ballon, the sensuous embraces, the turned-out feet, the tutus — all of dance's traditions were ridiculed. Under Nijinsky's direction, the audience saw only the dancers' profiles, their bodies hunched over, their heads hanging down, their turned-in feet hammering the wooden stage. The dancers later said the dance jarred their organs. It was a ballet as furiously new as the music.

The Parisian police eventually arrived. They only caused more chaos. Gertrude Stein described the scene: "We could hear nothing . . . The dancing was very fine and that we could see although our attention was constantly distracted by a man in the box next to us flourishing his cane, and finally, in a violent altercation with an enthusiast in the box next to him, his cane came down and smashed the opera hat the other had just put on in defiance. It was all incredibly fierce." The furor didn't end until the music stopped.

If there was any consolation from the violence that night, it was the publicity. Stravinsky's orchestral work was the talk of the town. He was suddenly cooler than Colette. Stravinsky would later remember the night as bittersweet. No one had heard his art, but he had become a genuine celebrity, the icon of the avant-garde. When the performance was over and the theater was empty, Diaghilev said only one thing to Stravinsky: "Exactly what I wanted."

Why did the crowd riot that night? How could a piece of music move a crowd to violence? This is *The Rite*'s secret. For the audience, Stravinsky's new work was the sound of remorseless originality. The crowd was expecting more Chopin. What they got instead was the gory birth of modern music.

The source of this pain is literally visible in the score. Stutterings of notes fill page after page. Densities of black, clots of sound. Pure, painful sonority, interrupted only by some spooky clarinet solo off in the distance. Even the instrumentation of *The Rite* insults the symphonic tradition. Stravinsky ignored the string instruments, the workhorse of the romantic composer. He found their fragile sound too much like a human voice. He wanted a symphonic sound without people, the sound of music "before the arising of Beauty."*

Stravinsky created this effect by conceding nothing to his audience. He disfigured its traditions and dismantled its illusions. While the crowd at the premiere assumed that beauty was immutable — some chords were just more pleasing than others — Stravinsky knew better. An instinctive modernist, he realized that our sense of prettiness is malleable, and that the harmonies we worship and the tonic chords we trust are not sacred. Nothing is sacred. Nature is noise. Music is nothing but a sliver of sound that we have learned how to hear. With *The Rite*, Stravinsky announced that it was time to learn something new.

This faith in our mind's plasticity — our ability to adapt to new

---

* How painful is *The Rite*? Zelda and F. Scott Fitzgerald used to make their dinner guests choose between listening to a scratchy recording of *The Rite* or looking at photographs of mutilated soldiers. Apparently, they thought the two experiences were roughly equivalent.

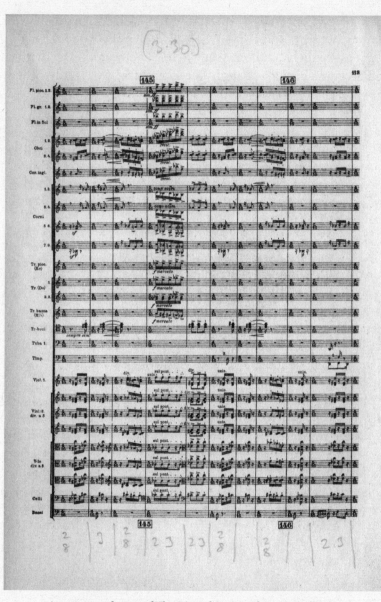

An annotated score of The Rite of Spring, *from the collection of composer Leopold Stokowski. Stokowski would later arrange the symphony for the Walt Disney cartoon* Fantasia.

kinds of music — was Stravinsky's enduring insight. When he was first composing *The Rite,* in Switzerland, testing out its dissonant chords on his piano, a young neighborhood boy got into the habit of yelling, "That's wrong!" at his window. Stravinsky just shrugged. He knew the brain would eventually right his wrongness. The audience would adapt to his difficult notes and discover the beauty locked inside his art. As neuroscience now knows, our sense of sound is a work in progress. Neurons in the auditory cortex are constantly being altered by the songs and symphonies we listen to. Nothing is difficult forever.

The Birth of Dissonance

Igor Stravinsky was born in 1882, the third son of minor nobles. His father was a St. Petersburg opera singer. Although his family insisted that he go to law school, Stravinsky hated law. The legal system embodied everything he found tedious: rules, forms, judges. Suffering through his classes, the young Igor steeped himself in angst. He would later describe his childhood "as a period of waiting for the moment when I could send everyone and everything connected with it to hell."

That moment arrived when his father died. Igor was now free to quit law school. He quickly joined the music academy of Nikolai Rimsky-Korsakov, the only important teacher he would ever have. Like the mature Stravinsky, who made modern symphonies by recycling old folk tunes, Korsakov was a composer defined by his contradictions. He was a Russian nationalist who loved German music, a czarist with a soft spot for the fin-de-siècle.

At the conservatory in St. Petersburg, Korsakov indoctrinated Igor into the anxiety of the modern composer. The problem facing modern music, Korsakov said, was simple: orchestral music had become boring. Wagner's vaunted ambition had been largely replaced by cheerful pastiche, most of it written for the ballet. (In his typical fashion, Wagner blamed this trend on the Jews.) Even more worrying, the modernist revolution seemed to be leaving composers behind. Painters were busy discovering abstraction, but music was already abstract. Poets were celebrating symbolism, but music had always been symbolic. Music could get no grander than *The Ring Cycle* and no more precise than Bach. The modern composer was trapped by the past. For this reason, the revolution in sound would have to begin with an act of deconstruction. As Wagner had declared half a century earlier when he embarked on his own violent renovation of musical style, "Works of art cannot be created at present, they can only be prepared for by means of revolutionary activity, by destroying and crushing everything that is worth destroying and crushing."*

The modernist coup d'état occurred in 1908, when Arnold Schoenberg decided to abandon the structure of classical music. As an act of aesthetic revolt, this was equivalent to a novelist abandoning plot. Before Schoenberg, every symphony followed a few simple rules. First, the composer introduced the tonic triad, a chord of three notes.† This chord was the invisible center of the music, the gravitational force that ordered its unfolding. Next, the composer carefully wandered away from the tonic triad, but never too far away. (The greater the acoustic distance from the tonic, the greater the dissonance, and too much dissonance was considered impolite.) The music always concluded with the tonic's triumphant return, the happy sound of a harmonic ending.

Schoenberg found this form suffocating. He wanted the structure of his music to reflect his own expressive needs and not the stuffy habits of some "mediocre kitschmonger." He began daydreaming of "the day when dissonance will be emancipated," when the symphony would be set free from the easy clichés of the eight-note

---

* It was often said at the time that the riot for *The Rite* was the worst musical riot since 1861, when an audience screamed at Wagner's *Tannhaüser*. After the debacle on his own opening night, Stravinsky consoled himself with this fact: "They hissed Wagner at forty-five years of age. I am only thirty-five. I too shall witness my triumph before I die" (Kelly, 299).

† A triad consists of a root note, a note that is a third above the root, and a note that is a third above that note. For example, the triad of C consist of the notes C-E-G. This follows directly from the octave of the C major scale, which is C-D-E-F-G-A-B-C.

scale. "If I must commit artistic suicide," Schoenberg said, "then I must live by it."

This suicide by atonality finally happened in the middle of Schoenberg's String Quartet no. 2 in F-sharp Minor, written in 1908. The quartet is a study of tonal entropy: we hear the slow decay of the F-sharp minor key. By the third movement of the string quartet, about the time a soprano begins to sing, "I am only a roaring of a prophetic voice," the tonal structure has been completely obliterated. No single harmonic endures for more than a few flirtatious seconds. The work as a whole is guided only by its parts. Classical music has been deconstructed.

In the program that night, Schoenberg tried to explain the logic behind his "pandemonium." He needed freedom from form because musical form had ceased to mean anything. "The overwhelming multitude of dissonances" could no longer be suppressed or censored. Schoenberg was finished following everyone else's rules. It was time to write his own.

When Stravinsky first heard Schoenberg, he immediately recognized the older musician's importance. A line had been crossed. The composer was now free to express *anything*, even ugly things. Stravinsky wrote in an early letter, "Schoenberg is one of the greatest creative spirits of our era."

The Viennese public did not agree. Even before Schoenberg completely abandoned tonality, his compositions stretched the limits of good taste. His riff on Wagner, *Verklärte Nacht (Transfigured Night)*, written in the 1890s, was banned by a Viennese musical society because it contained an unknown dissonance. The prim society didn't realize that Schoenberg was *only* interested in unknown sounds. If a dissonant sound was known, then it wasn't dissonant enough. Schoenberg liked to use chemistry as a metaphor for his music, a science in which tiny alterations can create an inordinately potent chemical. "One atom of hydrogen more, one less of carbon," Schoenberg wrote, "changes an uninteresting substance into a pigment or even an explosive."

By 1913 — the year of *The Rite* — Schoenberg had discovered how to make musical dynamite. His art wasn't just atonal; it was *painfully* atonal. During a concert performance of his Chamber Symphony no. 1, opus 9, just two months before the premiere of *The Rite*, the fragile relationship between the composer and his public finally imploded. Schoenberg's audience rebelled against his newness, screaming at the stage and forcing the police to be called and the program to be canceled.* Doctors declared, on behalf of traumatized listeners, that his atonality caused emotional and psychic distress. The tabloids were filled with headlines about lawsuits and fistfights. Schoenberg was unrepentant: "If it is art, it is not for all, and if it is for all, it is not art."

Petrushka was Stravinsky's first major work to follow in the brazen path of Schoenberg's avant-garde. But Stravinsky, unlike Schoenberg, did not undermine tonality by erasing it. He worried that atonality was too stifling, and that Schoenberg, with all his "rationalism and rules," might end up becoming "a dolled-up Brahms." Instead, Stravinsky decided to torment his audience by making it *overdose* on tonality. In *Petrushka,* a Diaghilev ballet about a puppet who comes to life, Stravinsky took two old folk melodies and set them against each other, like wind-up dolls. As a result, the music is bitonal, unfolding in two keys (F-sharp major, which is almost all black keys, and C major, which is all white keys) simultaneously. The result is unresolved ambiguity, the ironic dissonance of too much consonance. The ear must choose what to hear.

Waves of Noise

Our sense of sound begins when a sound wave, hurtling through space at 1,100 feet per second, collides with the eardrum. This shudder moves the tiniest three bones in the body, a skeleton locked in-

---

* They were also sounding off against Alban Berg's *Altenberglieder* and *Uber die Grenzen* (*Beyond the Bounds*), in which the twelve tones of the chromatic scale were sounded simultaneously. Berg was a student of Schoenberg, and Schoenberg was conducting the music that night.

side the ear, pressing them against the fluid-filled membranes of the cochlea. That fluid transforms the waves of compressed air into waves of salty liquid, which in turn move hair cells (so named because they look like microscopic bristles). This minute movement opens ion channels, causing the cells to swell with electricity. If the cells are bent at a sharp enough angle for long enough, they fire an electrical message onward to the brain. Silence is broken. Sound has begun.

The cochlea is quilted with 16,000 of these neurons. In a noisy world, they are ceaselessly being bent. The air is filled with vibrations, and every vibration reverberates inside the echo chamber of the ear. (Hair cells are sensitive to sounds of atomic dimensions. We can literally hear Brownian motion, the random jostle of atoms.) But how, out of this electrical cacophony, do we ever hear a coherent sound?

The answer is anatomical. Hair cells are arranged like the keys on a piano. On one end, they are tuned to respond to high-frequency sounds, while at the other end they are bent by the throb of low frequencies. When a scale is played, the hair cells mirror the escalating notes. They sway in time with the music, deftly translating the energy of noise into a spatial code of electricity.

And while every sound starts as a temporary pattern of hair cells, that's only the beginning of listening. In the time it takes to play a sixteenth note, the sensory rumors heard by the ear are rehearsed again and again inside the brain. Eventually, the sound reaches the primary auditory cortex, where neurons are designed to detect specific pitches. Instead of representing the full spectrum of sound waves vibrating inside the ear, the auditory cortex focuses on finding the note amid the noise. We tune out the cacophony we can't understand. (This is why we can recognize a single musical pitch played by different instruments. Although a trumpet and violin produce very different sound waves, we are designed to ignore these differences. All we care about is the pitch.) When these selective neurons in the auditory cortex become excited, the vague shudders of air finally become a musical note.

But a work of music is not simply a set of individual notes arranged in time. Music really begins when the separate pitches are melted into a pattern. This is a consequence of the brain's own limitations. Music is the pleasurable overflow of information. Whenever a noise exceeds our processing abilities — we can't decipher all the different sound waves hitting our hair cells — the mind surrenders. It stops trying to understand the individual notes and seeks instead to understand the relationships *between* the notes. The human auditory cortex pulls off this feat by using its short-term memory for sound (in the left posterior hemisphere) to uncover patterns at the larger level of the phrase, motif, and movement. This new approximation lets us extract *order* from all these notes haphazardly flying through space, and the brain is obsessed with order. We need our sensations to make sense.

It is this psychological instinct — this desperate neuronal search for a pattern, any pattern — that is the source of music. When we listen to a symphony, we hear a noise in motion, each note blurring into the next. The sound seems *continuous*. Of course, the physical reality is that each sound wave is really a separate thing, as discrete as the notes written in the score. But this isn't the way we experience the music. We continually abstract on our own inputs, inventing patterns in order to keep pace with the onrush of noise. And once the brain finds a pattern, it immediately starts to make predictions, imagining what notes will come next. It projects imaginary order into the future, transposing the melody we have just heard into the melody we expect. By listening for patterns, by interpreting every note in terms of expectations, we turn the scraps of sound into the ebb and flow of a symphony.

The Tension of Emotion

The structure of music reflects the human brain's penchant for patterns. Tonal music (that is, most baroque, classical, and romantic music) begins by establishing a melodic pattern by way of the tonic triad. This pattern establishes the key that will frame the song. The brain desperately needs this structure, as it gives it a way to organize the ensuing tumult of notes. A key or theme is stated in a mnemonic pattern, and then it is avoided, and then it returns, in a moment of consonant repose.

But before a pattern can be desired by the brain, that pattern must play hard to get. Music only excites us when it makes the auditory cortex struggle to uncover its order. If the music is too obvious, if its patterns are always present, it is annoyingly boring. This is why composers introduce the tonic note in the beginning of the song and then studiously avoid it until the end. The longer we are denied the pattern we expect, the greater the emotional release when the pattern returns, safe and sound. The auditory cortex rejoices. It has found the order it has been looking for.

To demonstrate this psychological principle, the musicologist Leonard Meyer, in his classic book *Emotion and Meaning in Music* (1956), analyzed the fifth movement of Beethoven's String Quartet in C-sharp Minor, opus 131. Meyer wanted to show how music is defined by its flirtation *with* — but not submission *to* — expectations of order. He dissected fifty measures of Beethoven's masterpiece, showing how Beethoven begins with the clear statement of a rhythmic and harmonic pattern and then, in an intricate tonal dance, carefully avoids repeating it. What Beethoven does instead is suggest variations of the pattern. He is its evasive shadow. If E major is the tonic, Beethoven plays incomplete versions of the E major chord, always careful to avoid its straight expression. He preserves an element of uncertainty in his music, making our brains beg for the one chord he refuses to give us. Beethoven saves that chord for the end.

According to Meyer, it is the suspenseful tension of music (arising out of our unfulfilled expectations) that is the source of the music's feeling. While earlier theories of music focused on the way a noise can refer to the real world of images and experiences (its connotative meaning), Meyer argued that the emotions we find in music come from the unfolding events of the music itself. This "embodied meaning" arises from the patterns the symphony invokes and then ignores, from the ambiguity it creates inside its own form. "For the human mind," Meyer wrote, "such states of doubt and confusion are abhorrent. When confronted with them, the mind attempts to resolve them into clarity and certainty." And so we wait, expectantly, for the resolution of E major, for Beethoven's established pattern to be completed. This nervous anticipation, says Meyer, "is the whole raison d'être of the passage, for its purpose is precisely to delay the cadence in the tonic." The uncertainty makes the feeling. Music is a form whose meaning depends upon its violation.

Stravinsky's music is *all* violation. While the cultured public thought music was just a collection of consonant chords played in neat meter, Stravinsky realized that they were wrong. Pretty noises are boring. Music is only interesting when it confronts us with tension, and the source of tension is conflict. Stravinsky's insight was that what the audience really wanted was to be denied what it wanted.

The Rite was the first symphonic work to express this antagonistic philosophy. Stravinsky anticipated the anticipations of his audience and then refused them every single one. He took the standard springtime song and made art out of its opposite. Dissonance never submits to consonance. Order does not defeat disorder. There is an obscene amount of tension, but it never gets resolved. Everything only gets worse. And then it ends.

To our sensitive nerves, such a musical work feels like an affront. The brain — an organ of synaptic habit — is hopelessly frustrated. We begin identifying with the violent sacrificial dance on stage. This was Stravinsky's intention: his music was a blatant provocation. Needless to say, it worked. But why? How did Stravinsky engineer so much agony into his art?

The answer to this question returns us to the idea of order. Although music begins with our predilection for patterns, the *feeling* of music, as Meyer noted, begins when the pattern we imagine starts to break down. *The Rite,* of course, is one long breakdown. Stravinsky didn't just invent some new musical patterns; he insisted on murdering our old ones. He introduced fragments of folk songs, then destroyed them with a gunshot of chromatic bullets. He took the big sonoric brushstrokes of major chords and put them through a cubist machine. Strauss is punked, Wagner is inverted, Chopin is mocked. Classicism is made cynical.

The sadistic newness of *The Rite*'s patterns, its stubborn refusal to conform to our learned expectations, is the dirty secret of its discontent. By disobeying every rule we think we know, Stravinsky forces us to confront the fact that we *have* expectations, that the mind anticipates certain types of order, followed by certain types of release. But in *The Rite,* these expectations are rendered useless. We do not know what note will come next. And this makes us angry.

The emotions generated by musical tension — a tension taken to grotesque heights by Stravinsky — throb throughout the body. As soon as the orchestra starts to play, the flesh undergoes a range of changes. The pupils dilate, pulse and blood pressure rise, the electrical conductance of the skin lowers, and the cerebellum, a brain region associated with bodily movement, becomes unusually active. Blood is even redirected to the leg muscles. (As a result, we begin tapping our feet in time with the beat.) Sound stirs us at our biological roots. As Schopenhauer wrote, "It is we ourselves who are tortured by the strings."

Stravinsky, of course, knew exactly how to raise our blood pressure. At first glance, this might seem like a dubious achievement. Must modern art really be so cruel? Whatever happened to beauty? But Stravinsky's malevolence was rooted in a deep understanding of the mind. He realized that the engine of music is conflict, not consonance. The art that makes us feel is the art that makes us hurt. And nothing hurts us like a pitiless symphony.

Why is music capable of inflicting such pain? Because it works on our feelings directly. No ideas interfere with its emotions. This is why "all art aspires to the condition of music." The symphony gives us the thrill of uncertainty — the pleasurable anxiety of searching for a pattern — but without the risks of real life. When we listen to music, we are moved by an abstraction. We feel, but we don't know why.

Stravinsky the Hipster

Stravinsky originally composed the "Augurs" section — the awful sound that started the riot — at a piano. One hand played the E major chord, and the other hand played the E-flat-major-seventh chord. He beat these sounds into the ivory and ebony keys, the rhythm as insistent as an alarm clock. To this headache he added some syncopated accents. The terror now had a groove. When Diaghilev first heard the "Augurs" section, he asked Stravinsky a single question: "How long will it go on like that?" Stravinsky's answer: "To the end, my dear." Diaghilev winced.

The irony is that the terrible beauty of the "Augurs" chord is not really dissonant. The sound is actually composed of classic tonal chords set against each other, in dissonant *conjunction*. Stravinsky melts together two separate harmonic poles, which has a short-circuiting effect. The ear hears shards of harmony (E, E-flat, C), but the brain can't fit the shards together. How come?

Because the sound is *new*. Stravinsky electrified the familiar. Never before in the history of music had a composer dared to confront us with this particular shiver of compressed air, played with this kind of naked staccato. The brain is befuddled, its cells baffled. We have no idea what this sound *is*, or where it might *go*, or what note will come *next*. We feel the tension, but we can't imagine the release. This is the shock of the new.

Stravinsky worshipped the new. "Very little tradition lies behind *Le Sacre du Printemps*," he would later brag. "And there is no theory. I had only my ear to help me." Stravinsky believed that music, like nature, required constant upheaval. If it wasn't original, then it wasn't interesting. As a result, Stravinsky spent his career constantly reinventing himself, divvying up his life into distinct stylistic periods. First, there is Stravinsky the modernist, the musical equivalent of Picasso. When that got boring, Stravinsky rebranded himself as a satirist of the baroque. From there, he ventured into minimalism, which morphed into neoclassicism, which became, at the end of his life, a Schoenbergian-sounding serialism. There was scarcely an *-ism* he hadn't explored.

Why didn't Stravinsky stay the same? As Bob Dylan, another musical chameleon, once remarked, "He not busy being born is busy dying." Stravinsky's greatest fear was dying the slow death of predictability. He wanted every one of his notes to vibrate with surprise, to keep the audience on edge. He believed that sheer daring — as opposed to beauty or truth — was "the motive force of the finest and greatest artist. I approve of daring: I set no limits to it."

Stravinsky's impudence manifested itself everywhere in his music. He never met a rhythm he didn't want to syncopate or a melody he wasn't compelled to mock. But Stravinsky's urge to unsettle was most evident in his avoidance of the musical patterns of the past. He knew that over time the dissonance of newness became consonant. The mind learned how to interpret the obscure noise. As a result, the symphony ceased to be scary.

Stravinsky himself had nothing against major chords or the charming patterns of the past. In fact, *The Rite* is full of consonant pitches and allusions to old Russian folk tunes. But Stravinsky needed tension. He wanted his music to seethe with stress, with the Sturm und Drang of originality. And the only way to create that kind of music was to create a new kind of music, a dissonant sound that the audience didn't know. "I have confided to my orchestra the great fear which weighs on every sensitive soul confronted with *potentialities*," Stravinsky wrote in an introduction to *The Rite*. It was this sound — the fearful sound of newness without limits, a form that could potentially become *any* form — that Stravinsky heard in the pubescent stirrings of spring.

The problematic balancing act, of course, is being au courant without being incoherent, difficult but not impossible. Although *The Rite* was described at its premiere as an example of pure noise, chaos without cessation, it is actually an intricate quilt of patterns. Stravinsky was nothing if not meticulous. But the patterns he wove into *The Rite* weren't the usual patterns of Western music. The brain didn't know these patterns. Stravinsky abjured all conventions, and, in this symphonic work, created his own system of harmony and rhythm. Take the beat of the "Augurs" chord, a rhythm that is usually described as arbitrary and random. It is actually anything but. Stravinsky uses this moment to suspend melody and harmony, so that all we hear are the dissonant stabs. He wants us to focus on the source of our pain.

And while the berating beat of the section pretends to appear abruptly, Stravinsky has carefully prepared us for its rhythmic throb. Amid the swarming folk melodies of the introduction, Stravinsky has the violins outline the chord (D-flat, B-flat, E-flat, B-flat) and the beat (2/4 time), so that when the actual "Augurs" begins we hear their stuttering accents as either on or off the beat. Their jarring randomness is framed, which (ironically) only serves to emphasize their randomness. And while Stravinsky is busy destabilizing the beat, using the "Augurs" chord to dismantle the pattern he has just created, he is simultaneously making us — the audience — keep the beat steady. He forces us to project his own peculiar pattern forward in time. In fact, we quickly realize that we are the *only* source of steadiness: the helpful violins have gone silent. Stravinsky has used our addiction to patterns to give his disorder an order. The order is our own.

This is the method of *The Rite*. First, Stravinsky throws a wrench into our pattern-making process, deliberately and loudly subverting everything we think we know. (This is the not-so-subtle purpose of the "Augurs" chord.) Then, after clearing our heads of classical detritus, Stravinsky forces us to generate patterns from the music itself, and not from our preconceived notions of what the music *should* be like. By abandoning the conventions of the past, he leaves us with no pattern but that which we find inside his own ballet music. If we try to impose an outside pattern onto *The Rite*, if we try to unlock its newness using the patterns of Beethoven or Wagner or *Petrushka*, those patterns will be hurled back in our faces. Even when we can recognize Stravinsky's notes, their arrangement confuses us, for Stravinsky fragments everything. His imagination was a blender.

All of this novelty leaves us bitterly disoriented. To find the echo of order in *The Rite,* we have to pay exquisite attention. If we fail to listen carefully, if we tune out its engineered undulations, then the whole orchestra becomes nothing more than a mutiny of noise. The music disappears. This is what Stravinsky wanted. "To listen is an effort," he once said, "and just to hear is no merit. A duck hears also."

But Stravinsky makes it difficult to listen. His orchestral work is a stampede, and whatever fragile order there is remains hidden. We hear patterns, but barely. We sense a structure, but it seems to exist only in our minds. Is this music? we wonder. Or is this noise? Are these quavers random, or is there a method to their madness? Stravinsky doesn't stop to answer these questions. He doesn't even stop to acknowledge that a question has been asked. The sound just stampedes on.

The sonic result is pure ambiguity, on a terrifying scale. When Stravinsky said that "Music is, by its very nature, essentially powerless to express anything at all," he was alluding to the fact that the epitome of musical expression is uncertainty. If music is not ambiguous then it expresses nothing, and if it expresses something then it has only expressed the absence of certainty. And while many composers before *The Rite* made a habit of limiting originality — too much newness was too painful — Stravinsky pitilessly ended that aesthetic game. His symphonic music denies us a consonant climax. It mocks our expectations of a happy ending. In fact, it mocks all our expectations.

And so, at the exact moment when every other composer in every other symphony is contemplating the angle of easy repose — the satisfied sound of an orchestra ending with the tonic — Stravinsky decides to kill his virgin with some big timpani drums. He forces her to dance the impossible dance, giving her a different beat in every musical bar. The rhythmic patterns fly by in a schizophrenic babble: 9/8 becomes 5/8, which becomes 3/8, which abruptly shifts into 2/4–7/4–3/4, and so on. Our cells can sense the chaos here; we know that this particular wall of sound is irresolvable. All we can do is wait. This too must end.

Plato's Mistake

What *is* music? This is the unanswered question at the center of *The Rite*. The violent crowd at the premiere insisted that the symphonic work was nothing but noise. Whatever music was, this wasn't it. There were limits to newness, and *The Rite* had crossed the line.

Stravinsky, of course, believed otherwise. He said that noise became music "only by being organized, and that such an organization presupposes a conscious human act." Music, *The Rite of Spring* shouts, is man-made, a collection of noises that we have learned how to hear. That is all.

This was a radically new definition of music. Ever since Plato, music had been seen as a metaphor for the innate order of nature. We don't make music, Plato said, we *find* it. While reality appears noisy, hidden in the noise is an essential harmony, "a gift of the Muses." For Plato, this made music a form of medicine, "an ally in the fight to bring order to any orbit in our souls that has become unharmonized." The beauty of the C major chord reflected its rational trembling, which could inspire a parallel rationality inside us.

Plato took the power of art seriously. He insisted that music (along with poetry and drama) be strictly censored inside his imaginary republic. Seduced by the numerical mysticism of Pythagoras, Plato believed that only consonant musical pitches — since they vibrated in neat geometrical ratios — were conducive to rational thinking, which is when "the passions work at the direction of reason." Unfortunately, this meant systematically silencing all dissonant notes and patterns, since dissonance unsettled the soul. Feelings were dangerous.

At first glance, *The Rite of Spring* seems like perfect evidence for Plato's theory of music. Stravinsky's orchestral dissonance provoked a violent urban riot. This is exactly why the avant-garde must be banned: it's bad for the republic. Better to loop some easy elevator music.

But Plato — for all of his utopian insight — misunderstood what music actually is. Music is *only* feeling. It *always* upsets our soul. If we censored every song that filled people with irrational emotions, then we would have no songs left to play. And while Plato only trusted those notes that obeyed his mathematical definition of order, music really begins when that order collapses. We make art out of the uncertainty.

The Rite shattered many myths, but the myth Stravinsky took the most pleasure in shattering was the parable of progress. Stravinsky said, "In music, advance is only in the sense of developing the instruments of the language." While Plato believed that music would one day perfectly mirror the harmony of the cosmos — and thus inspire our souls with the pure sound of reason — Stravinsky's symphonies were monuments to the meaninglessness of progress. In the modernist vision of *The Rite*, music is simply a syntax of violated patterns. It doesn't become better over time, it just becomes different.

Stravinsky's version of progress was borne out by what happened after the riot. Although that first audience screamed at the stage, cursing the ballet's bitter abandonment of every known tradition, *The Rite* went on to define its own tradition. In fact, within a few years of its premiere, *The Rite* was being performed to standing ovations and Stravinsky was being carried out of the auditorium on the shoulders of the crowd. Diaghilev joked, "Our little Igor now needs police escorts out of his concerts, like a prizefighter." The same symphony that once caused a violent riot became the clichéd example of modern music. Audiences were able to hear its delicate patterns and found the frightening beauty buried in its undulations. By 1940, Walt Disney used *The Rite* in the sound track of *Fantasia*. The "Augurs" chord was fit for a cartoon.

The stubborn endurance of *The Rite* was its most subversive triumph. If Platonists believed that music had a natural definition, its order a reflection of some mathematical order outside of us, Stravinsky's symphony forced us to admit that music is our own creation. There is nothing holy about the symphony: it is simply some vibrating air that our brains have *learned* how to hear.

But how do we learn how to hear music? How does an oblivion of noise become a classic modern symphony? How does the pain of *The Rite* become pleasurable? The answer to these questions returns us to the brain's unique talent: its ability to change itself. The auditory cortex, like all our sensory areas, is deeply plastic. Neuroscience, stealing vocabulary from music, has named these malleable cells the cortico*fugal* network, after the fugal form Bach made famous. These contrapuntal neurons feed back to the very substrate of hearing, altering the specific frequencies, amplitudes, and timing patterns that sensory cells actually respond to. The brain tunes its own sense of sound, just as violinists tune the strings of their instruments.

One of the central functions of the corticofugal network is what neuroscience calls egocentric selection. When a pattern of noises is heard repeatedly, the brain memorizes that pattern. Feedback from higher-up brain regions reorganizes the auditory cortex, which makes it easier to hear the pattern in the future. This learning is largely the handiwork of dopamine, which modulates the cellular mechanisms underlying plasticity.

But what orders the corticofugal feedback? Who is in charge of our sensations? The answer is experience. While human nature largely determines how we hear the *notes,* it is nurture that lets us hear the *music.* From the three-minute pop song to the five-hour Wagner opera, the creations of our culture teach us to expect certain musical patterns, which over time are wired into our brain.

And once we learn these simple patterns, we become exquisitely sensitive to their variations. The brain is designed to learn by association: if this, then that. Music works by subtly toying with our expected associations, enticing us to make predictions and then confronting us with our prediction errors. In fact, the brainstem contains a network of neurons that responds *only* to surprising sounds. When the musical pattern we know is violated, these cells begin the neural process that ends with the release of dopamine, the same neurotransmitter that reorganizes the auditory cortex. (Dopamine is also the chemical source of our most intense emotions, which helps to explain the strange emotional power of music, especially when it confronts us with newness and dissonance.) By tempting us with fragile patterns, music taps into the most basic brain circuitry.

But dopamine has a dark side. When the dopamine system is imbalanced, the result is schizophrenia.* If dopamine neurons can't correlate their firing with outside events, the brain is unable to make cogent associations. Schizophrenics have elaborate auditory hallucinations precisely because their sensations do not match their mental predictions. As a result, they invent patterns where there are none and can't see the patterns that actually do exist.

The premiere of *The Rite,* with its methodical dismantling of the audience's musical expectations, literally simulated madness. By subverting the listeners' dopamine neurons, it also subverted their sanity. Everything about it felt wrong. Pierre Monteux, the conductor, said he was convinced that Stravinsky was a lunatic. During the symphony's lengthy rehearsals — it required twice as many sessions as *The Firebird* — the violinists denounced *The Rite* as *"schmutzig"* (dirty). Puccini said it was "the work of a madman." The brass musicians playing the massive fortissimos broke into spontaneous fits of laughter. Stravinsky took the long view. "Gentlemen, you do not have to laugh," he drolly told the rehearsing orchestra. "I know what I wrote."

With time, the musicians came to understand Stravinsky's method. His creativity was seared into their brains as their dopamine neu-

* While schizophrenia cannot be reduced to any single anatomical cause, the dopaminergic hypothesis is neuroscience's most tenable explanation. According to this theory, many of the symptoms of schizophrenia are caused by an excess of certain dopamine receptor subtypes, especially in the mesolimbic-mesocortical dopamine system.

rons adjusted to the "Augurs" chord. What once seemed like a void of noise became an expression of difficult magnificence. This is the corticofugal system at work. It takes a dissonant sound, a pattern we can't comprehend, and makes it comprehensible. As a result, the pain of *The Rite* becomes bearable. And then it becomes beautiful.

The corticofugal system has one very interesting side effect. Although it evolved to expand our minds — letting us learn an infinitude of new patterns — it can also limit our experiences. This is because the corticofugal system is a positive-feedback loop, a system whose output causes its input to recur. Think of the microphone placed too close to the speaker, so that the microphone amplifies its own sound. The resulting loop is a meaningless screech of white noise, the sound of uninterrupted positive feedback. Over time, the auditory cortex works the same way; we become better able to hear those sounds that we have heard before. This only encourages us to listen to the golden oldies we already know (since they sound better), and to ignore the difficult songs that we don't know (since they sound harsh and noisy, and release unpleasant amounts of dopamine). We are built to abhor the uncertainty of newness.

How do we escape this neurological trap? By paying attention to art. The artist is engaged in a perpetual struggle against the positive-feedback loop of the brain, desperate to create an experience that no one has ever had before. And while the poet must struggle to invent a new metaphor and the novelist a new story, the composer must discover the undiscovered pattern, for the originality is the source of the emotion. If the art feels difficult, it is only because our neurons are stretching to understand it. The pain flows from the growth. As Nietzsche sadistically declared, "If something is to stay in the memory it must be burned in; only that which never ceases to hurt stays in the memory."

This newness, however torturous, is necessary. Positive-feedback loops, like that shrieking microphone, always devour themselves. Without artists like Stravinsky who compulsively make everything new, our sense of sound would become increasingly narrow. Music would lose its essential uncertainty. Dopamine would cease to flow. As a result, the feeling would be slowly drained out of the notes, and all we would be left with would be a shell of easy consonance, the polite drivel of perfectly predictable music. Works like *The Rite** jolt us out of this complacency. They keep us literally open-minded. If not for the difficulty of the avant-garde, we would worship nothing but that which we already know.

This is what Stravinsky knew: music is made by the mind, and the mind can learn to listen to almost anything. Given time, even the intransigent *Rite* would become just another musical classic, numbing listeners with its beauty. Its strange patterns would be memorized, and they would cease to hurt. The knifing chord of the "Augurs" would become dull with use, and all the meticulously engineered dissonances would fade into a tepid kind of pulchritude. This was Stravinsky's nightmare, and he knew it would come true.

What separated Stravinsky from his rioting audience that night was his belief in the limitless possibilities of the mind. Because our human brain can learn to listen to anything, music has no cage. All music needs is a violated pattern, an order interrupted by a disorder, for in that acoustic friction, we hallucinate a feeling. Music is that feeling. *The Rite of Spring* was the first symphonic work to celebrate this fact. It is the sound of art changing the brain.

* Or artists like Bob Dylan, who took folk music electric, or the Ramones, who punked out rock music . . . Musical history is largely the story of artists who dared to challenge the expectations of their audiences.

Chapter 7

Gertrude Stein

The Structure of Language

> Words are finite organs of the infinite mind.
>
> — Ralph Waldo Emerson

BEFORE GERTRUDE STEIN was an avant-garde artist, she was a scientist. Her first published piece of writing was in the May 1898 edition of the *Psychological Review*.* The article summarized her research in the Harvard psychology lab of William James, where Stein was exploring automatic writing.† In her experiments, she used a wooden planchette — a device normally used to try to contact the dead — in order to channel her own subconscious. Stein wanted to write down whatever words first entered her mind.

The result was predictably ridiculous. Instead of revealing the mind's repressed interiors, Stein's automatic-writing experiments generated a lot of spontaneous gibberish. She filled page after page with inscrutable sentences like this: "When he could not be the longest and thus to be, and thus to be, the strongest." What could these

---

* Stein was a coauthor with Leo Solomons on an earlier science article, "Normal Motor Automatism," published in September 1896. However, she probably had little role in writing this paper.

† By the time her science article appeared in print, Stein was on her way to medical school at Johns Hopkins. At Hopkins, Stein would work in the neuroanatomy lab of Franklin Mall, a leading brain anatomist.

words possibly mean? After analyzing the data, Stein concluded that they didn't mean anything. Her experiment hadn't worked. "There are automatic movements but not automatic writings," she lamented. "Writing for the normal person is too complicated an activity to be indulged in automatically."

But Stein's experimental failure got her thinking. Even when she wrote about absolutely nothing, which was most of the time, her nothingness remained grammatical. The sentences were all meaningless, and yet they still obeyed the standard rules of syntax. Subjects matched verbs, adjectives modified nouns, and everything was in the right tense. "There is no good nonsense without sense," Stein concluded, "and so there cannot be automatic writing." Although she had hoped that her experiment would free language from its constraints, what she ended up discovering was the constraint that can't be escaped. Our language has a structure, and that structure is built into the brain.

It would be another decade before Stein converted her experimental conclusions into a new form of literature. Even Stein later admitted that her surreal writing style emerged from her automatic-writing experiments. The sentences she wrote in the laboratory inspired her lifelong obsession with words and rules, with how language works and why it's so essential to the human mind. Her art was born of her science.

Tender Buttons, written in 1912 but not published until 1914, was the first of Stein's books to attract widespread critical attention. (Her first book, *Three Lives*, sold only seventy-three copies.) *Tender Buttons* is divided into three arbitrary sections, "Objects," "Food," and "Rooms." "Objects" and "Food" are composed of short, epigrammatic pieces with titles like "Mutton" and "An Umbrella." But these objects are not Stein's subject. Her subject is language itself. The purpose of her prose poems, she said, was "to work on grammar and eliminate sound and sense." Instead of a plot, she gave us a lesson in linguistics.

Stein, as usual, advertised her audacity. The very first page of *Tender Buttons* serves as a warning: this is not a nineteenth-century novel. In place of the customary scene setting, or some telling glimpse of the main character, the book begins with an awkward metaphor:

A CARAFE, THAT IS A BLIND GLASS.

A kind in glass and a cousin, a spectacle and nothing strange a single hurt color and an arrangement in a system to pointing. All this and not ordinary, not unordered in not resembling.

This tricky paragraph is about the trickiness of language. Although we pretend our words are transparent — like a layer of glass through which we see the world — they are actually opaque. (The glass is "blind.") Stein is trying to remind us that our nouns, adjectives, and verbs are not real. They are just arbitrary signifiers, random conglomerations of syllables and sound. A *rose,* after all, is not really a rose. Its letters don't have thorns or perfumed petals.

Why, then, do we invest words with so much meaning? Why do we never notice their phoniness? Stein's revelation, which she had for the first time during her science experiments, was that everything we say is enclosed by "an arrangement in a system." This linguistic system, although invisible, keeps words from being "not unordered in not resembling." Because we instinctively "arrange" language, it seems like "nothing strange." Stein wanted us to acknowledge these hidden grammars, for it is their structure that makes language so meaningful and useful.

But if Stein wanted to talk about grammar, then why didn't she just talk about grammar? Why did she have to make everything so difficult? The answer to these questions can be found in the form of another psychological experiment, one that William James liked to use on unsuspecting undergraduates. He describes the essence of the experiment in his *Principles of Psychology,* suggesting a method whereby a mind can be made aware of the structure underneath our words: "If an unusual foreign word be introduced, if the grammar trips, or if a term from an incongruous vocabulary suddenly appears, the sentence detonates, as it were, we receive a shock from the incongruity, and the drowsy assent is gone."

Reading *Tender Buttons,* with its "grammar trips" and "incongruous vocabularies," is often an experiment in frustration. But this is precisely Stein's point. She wants us to feel the strictures of the sentence, to question our own mental habits. If nothing else, she wants to rid us of our "drowsy assent," to show us that language is not as simple as it seems. And so she fills her sentences with long sequences of non sequiturs. She repeats herself, and then she repeats her repetitions. She writes sentences in which her subjects have no verb, and sentences in which her verbs have no subject.

But the secret of Stein's difficulty is that it doesn't drive us away. Rather, it brings us in. Her words demand a closeness: to steal sense from them, we have to climb into them. This forced intimacy is what Stein wanted most of all, since it makes us question how language actually works. When suffering through her sentences, we become aware, she said, of "the way sentences diagram themselves," of the instinctive nature of syntax. Stein's inscrutability returns us to the grammar lessons of elementary school, when we first realized that the sentence is not simply the sum of its words. "Other things may be more exciting to others when they are at school," Stein wrote, "but to me undoubtedly when I was at school the really exciting thing was diagramming sentences." In her writing, Stein wanted to share the thrill.

It would take psychology nearly fifty years before it rediscovered the linguistic structures that Stein's writing had so assiduously exposed. In 1956, a shy linguist named Noam Chomsky announced that Stein was right: our words are bound by an invisible grammar, which is embedded in the brain. These deep structures are the secret sources of our sentences; their abstract rules order everything we say. By allowing us to combine words into meaningful sequences, they inspire the infinite possibilities of language. As Charles Darwin declared, "Language is an instinctive tendency to acquire an art." The genius of Stein's art was to show us how our language instinct works.

Picasso's Portrait

After finishing her automatic-writing experiments in William James's lab, Stein started medical school at Johns Hopkins. She spent her first two years dissecting the brains of embryos, chronicling the intricate development of the nervous system. She learned how to cut away the cortex and preserve the tissue in toxic vats of formaldehyde. When Stein wasn't in the lab, she enjoyed boxing and smoking cigars. Everyone said she was an excellent scientist.

Things began to unravel when Stein started her clinical rotations. "Practical medicine did not interest her," Stein would later write. She confessed that "she was bored, frankly openly bored." Instead of studying organic chemistry or memorizing her anatomy lessons, Stein stayed up late reading Henry James. She was inspired by the first murmurings of modernism and started making her own medical notes notoriously inscrutable.* As one professor remarked, "Either I am crazy or Miss Stein is."

In 1903, just a semester away from graduation, Stein moved to Paris. She settled in with her brother Leo, who had an apartment at 27 rue de Fleurus. Leo had just purchased his first Cézanne painting — "Anyone can buy paintings in Paris," he told Gertrude — and was beginning to enmesh himself in the local art scene. Gertrude made herself right at home. As she wrote in *Everybody's Autobiography,* "I joined him and sat down in there and pretty soon I was writing."

---

* In a typed letter to a professor about her anatomical drawings of the brainstem, Stein filled her prose with typical eccentricities and "errors": "They [the drawings] clear awaythe underbrush and leave a clear road. I had so muchdifficulty in understanding the conditions from the text books that I felt such a clarifming [clarifying] process to be much needed. N o t that XXXXX the books do not all tell the truth as I know it but that they tell so XXXXXXX much that one is confused . . ."

Her early work was influenced by the artists who lingered around the apartment. *Three Lives* was inspired by a Cézanne portrait. Her next book, *The Making of Americans,* emerged from her relationship with Matisse. But Stein was closest to Pablo Picasso. As she wrote in her essay *Picasso* (1938), "I was alone at this time in understanding him because I was expressing the same thing in literature."

Their relationship began in the spring of 1905, just as Picasso was becoming bored with his blue period. Gertrude Stein asked him to paint her portrait. The painter couldn't say no; not only had Stein's Saturday-night salons become a magnet for the Parisian avant-garde (Matisse, Braque, and Gris were normally there), but Gertrude and her brother Leo were some of his earliest benefactors. Their walls were lined with his experiments.

Picasso struggled with Stein's portrait as he had never struggled with another painting. Day after day, Stein returned to Picasso's apartment high in the hills of Montmartre. They talked while Picasso carefully reworked the paint on the canvas. They discussed art and philosophy, William James's psychology, Einstein's physics, and the gossip of the avant-garde. In Stein's autobiography — mischievously titled *The Autobiography of Alice B. Toklas* — she described the making of the picture:

> Picasso had never had anybody pose for him since he was sixteen years old. He was then twenty-four and Gertrude had never thought of having her portrait painted, and they do not know either of them how it came about. Anyway, it did, and she posed for this portrait ninety times and a great deal happened during that time . . . There was a large broken armchair where Gertrude Stein posed. There was a little kitchen chair where Picasso sat to paint, there was a large easel and there were many very large canvases. She took her pose, Picasso sat very tight in his chair and very close to his canvas and on a very small palette, which was of a brown gray color, mixed some more brown gray and the painting began. All of a sudden one day Picasso painted out the whole head. I can't see you anymore when I look, he said irritably, and so the picture was left like that.

Portrait of Gertrude Stein, *by Pablo Picasso, 1906*

But the picture wasn't left like that. Stein, writing from the imagined perspective of her lover Alice B. Toklas, is a reliably unreliable narrator. Picasso actually completed the head after a trip to Spain in the fall of 1906. What it was he saw there — ancient Iberian art or the weathered faces of peasants — has been debated, but his style changed forever. When he returned to Paris, he immediately began to rework Stein's portrait, giving her the complexion of a primitive mask. The perspective of her head was flattened out, and the painting became even more similar to Cézanne's painting of his wife, which Picasso had seen in Stein's apartment. When someone commented that Stein did not look like her portrait, Picasso replied, "She will."

Picasso was right. After he painted Stein's face, she began writing in an increasingly abstract style. Just as Picasso had experimented with painting — his art was now about the eloquence of incoherence — Stein wanted to separate language from the yoke of "having to say something." Modern literature, she announced, must admit its limits. Nothing can ever really be described. Words, like paint, are not a mirror.

By the time Stein started writing *Tender Buttons* a few years later, her chutzpah exceeded even Picasso's. Her modernist prose featured one jarring misnomer after another. "The care with which there is incredible justice and likeness," she writes almost intelligibly, "all this makes a magnificent asparagus." For Stein, making sense was just a comic setup; the punch line was the absurd asparagus.

As *Tender Buttons* progresses, this drift toward silliness is increasingly exaggerated. When Stein defines *dinner* toward the end of the work, her sentences have become little more than units of sound, a modernist "Jabberwocky": "Egg ear nuts, look a bout. Shoulder. Let it strange, sold in bell next herds." No dictionary will help you understand this literary disarray. In fact, using a dictionary will just make things worse.

This is because the meaning of Stein's abstract sentences — to the extent there is any meaning — depends entirely on the *un*reality of her words. By putting her words into ridiculous new arrangements, Stein forces us to see them anew, to "read without remembering." If "Egg ear nuts" is interesting, it is only because we have stopped understanding it one word at a time. An *egg* is no longer an egg. For Stein's writing to succeed, the sentence must become more than the sum of its separate definitions. There must be something else there, some mysterious structure that transcends her individual words. That something else is what makes *Tender Buttons* poetry and not just badly written prose.

And even though Stein's art was mocked and ridiculed and she had to pay her publisher to print her first book,* she never doubted her genius. At dinner parties, she liked to compare herself to Jesus and Shakespeare. Her art was difficult, she said, because it was so original, because no one had ever dared to write like her before. But hadn't Stravinsky survived the riot? Wasn't Cézanne now a cause célèbre? Didn't Jesus prove popular in the end? As Stein would later confide, "To see the things in a new way that is really difficult, everything prevents one, habits, schools, daily life, reason, necessities of daily life, indolence, everything prevents one, in fact there are very few geniuses in the world."

The James Brothers

Stein always wrote at night. With the streets of Paris quiet, she could ignore everything but herself, "struggling with the sentence, those long sentences that had to be so exactly carried out." She would compose her words in pencil on scraps of paper before correcting her composition, and copying her prose into the certainty of ink. Some nights she would write furiously fast, scribbling a page every two and half minutes. And then there were those endless nights when she couldn't write at all, and would just stare numbly at the blank page. But Stein sat at her desk anyway, stubbornly waiting for the silence to disappear. She only stopped working in the moments "before the dawn was clear," for light made things too real, too painfully distinct in their "thingness." The envelope of dark allowed Stein to ignore these distractions, focusing instead on the process of composition, the way her writing wrote itself. She would then sleep until the early afternoon.

Stein wasn't the first writer to disconnect her sentences from re-

---

* After reading *Three Lives*, Stein's publisher assumed that she was not fluent in English. He referred Stein to a copyeditor in Paris who could "fix" her manuscript. Stein, of course, refused to have her grammatical errors corrected. Her editor was not amused. "You have written a very peculiar book," he admonished, "and it will be a hard thing to make people take it seriously."

ality. Before Gertrude Stein converted cubism into a literary form, Henry James, William's younger brother, made a career out of writing famously verbose and ambiguous fiction. Nothing in James's later novels is described straight or directly. Instead, his prose constantly questions its own meaning. Everything is circumscribed by words, words, and more words, until the original object vanishes in a vapor of adjectives, modifiers, and subclauses. The world is swallowed by style.

It should come as no surprise that Gertrude Stein loved Henry's literature. She returned to his novels again and again, finding inspiration in his murky palimpsest of words. "You see," Stein once said of Henry James, "he made it sort of like an atmosphere, and it was not the realism of the characters but the realism of the composition which was the important thing." As she notes in *The Autobiography of Alice B. Toklas,* "Henry James was the first person in literature to find the way to the literary method of the twentieth century." She called him "the precursor."

Why did Gertrude define Henry's late fiction as the start of modern literature? Because he was the first writer to deprive the reader of the illusion that language directly reflects reality. In his novels, words are vague symbols that require careful interpretation. As a result, the meaning of every one of his sentences emerges not from the text alone, but from the interaction of the subjective reader and the unknowable work. A perfect truth or final reading always eludes our grasp, for reality, Henry wrote, "has not one window but a million . . . At each of them stands a figure with a pair of eyes."

Henry's literary philosophy reflected William's psychology. In his 1890 textbook *The Principles of Psychology,* William declared that "language works against our perception of the truth." Words make reality seem as if it is composed of discrete parts — like adjectives, nouns, and verbs — when in actual experience, all these different parts run together. William liked to remind his readers that the world is a "big blooming buzzing confusion," and that the neat concepts and categories we impose on our sensations are imaginary. As he wrote in his *Principles,* "It is, in short, the reinstatement of the vague to its proper place in our mental life which I am so anxious to press on the attention." At the time, critics quipped that William wrote psychology like a novelist, and Henry wrote novels like a psychologist.

Unfortunately, by the time Stein began writing in the Parisian darkness, modern literature and modern psychology had parted ways. While modernist writers pursued Henry's experiments, becoming ever more skeptical of the self and its sentences, modern psychology turned its back on William's view of the mind. A "New Psychology" had been born, and this rigorous science had no need for Jamesian vagueness. Measurement was now in vogue. Psychologists were busy trying to calculate all sorts of inane things, such as the time it takes for a single sensation to travel from your finger to your head. By quantifying consciousness, they hoped to make the mind fit for science.

William didn't think much of this New Psychology. He believed that its reductionist approach had lost touch with what reality *felt* like, privileging the mechanics of the brain over the "infinite inward iridescences" of the conscious mind. The fashionable obsession with measuring human sensations neglected the fact that every sensation was perceived as part of a whole process of thinking. (As William wrote, "Nobody ever had a sensation by itself.") To prove his point, William used language as a metaphor: "We ought to say," he wrote, "a feeling of *and,* a feeling of *if,* a feeling of *but,* and a feeling of *by,* quite as readily as we say a feeling of blue, or a feeling of cold."* Just as we ordinarily ignore the connecting words of sentences, focusing instead on their "substantive parts," the New Psychologists ignored the "transitive processes" at work in the mind. This was their crucial mistake. Sentences need articles and adverbs,

---

* Stein would later turn James's idea into a work of art. In her long and virtually impregnable poem "Patriarchal Poetry," Stein tried to call attention to all the parts of language that we usually neglect. And so she wrote a stanza all about adverbs: "Able able nearly nearly nearly nearly able able finally nearly able nearly not now finally finally nearly able." And a stanza about prepositions: "Put it with it with it and it and it in it in it add it add it at it at it with it with it put it put it to this to understand."

and the mind needs thoughts to connect its other thoughts together.

Stein would never forget William's philosophy. She told Richard Wright that "William James taught me all I know." He remained her hero until the end of her life: "A great deal I owe to a great teacher, William James. He said, 'Nothing has been proved.'" James once visited Stein's apartment in Paris and saw her walls full of Cézannes, Matisses, and Picassos. As Stein described it, "He looked and gasped, I told you, he said, I always told you that you should keep your mind open."

Stein's writing was a unique amalgamation of William's psychology and Henry's literature. Like the James brothers, she realized that sentences are practical abstractions. We give the world clarity by giving it names. Unfortunately, our names are pretend. When, in *Tender Buttons*, Stein asks, "What is cloudiness, is it a lining, is it a roll, is it a melting," she is questioning what *cloudiness* actually means.* After all, clouds are evanescent wisps, and no two are identical. So how can the same word describe such different things? Whereas realist writers before Stein had tried to pretend that our words neatly map onto the world, Stein called attention to the fact that words are subjective and symbolic. They are tools, not mirrors. As Ludwig Wittgenstein once said, "The meaning of a word is its use in the language." In her pugnacious prose, Stein tried to make words completely useless. She wanted to see what parts of language remained when words meant nothing.

The New Psychology that William James fiercely resisted didn't last very long. The brain failed to disclose its subtle secrets, and psychology grew bored with measuring the speed of the nervous system. By the 1920s, scientists were busy trying to empirically explain the mind from the *outside;* the brain had become a black box. This new approach was called behaviorism. According to the behaviorists, behavior was everything. There was no idea, belief, or emotion

* As Shakespeare observed in *A Midsummer Night's Dream,* we "give to airy nothing / A local habituation and a *name.*"

that couldn't be restated in terms of our actions. The world consisted of stimuli to which we responded, like animatronic machines.

The experimental evidence for this grand hypothesis was based on the work of two psychologists: Ivan Pavlov and Edward Thorndike. Pavlov worked in Russia, and Thorndike worked at Columbia University. Both demonstrated that rats, cats, and dogs were eminently trainable. Using positive reinforcement (a few morsels of food), Pavlov and Thorndike conditioned their hungry animals to do all sorts of stupid tricks. Rats would endlessly press levers, dogs would drool at the sound of a bell, and cats could learn how to escape labyrinthine mazes. The behaviorists believed they had explained the process of learning.

It didn't take long before scientists applied the behaviorist logic to humans. A rat was a dog was a person. With time, the human brain came to be seen as a reflex machine par excellence, an organ exquisitely sensitive to stimuli and response. In this reductionist framework, the mind was nothing but a network of conditioned instincts. We were completely free to be entrapped by our environment.

Obviously, this new approach to human nature begged a lot of questions. One of the first concerned language. How do children learn so many words and grammatical rules? (A two-year-old, for example, learns a new word every two hours.) Though we are born without language, within a few years our brains become *obsessed* with language. But how does this happen? How does a complicated and convoluted system of symbols hijack the mind?

The behaviorists weren't stymied for long. By the 1940s, they had explained away language as yet another creation of stimuli and response. According to behaviorist theory, parents provided the formative feedback that led children to correctly conjugate verbs, add on plurals, and pronounce words. Infants began by associating things with sounds, and then, over time, learned to combine those sounds into sentences. If someone looked at a red rose and said, "Red," that "verbal behavior" was merely a reflex prompted by the stimulus of redness, which his parents had taught him to associate with the proper adjective. If he said, "Rose," that was merely summarizing the collection of stimuli underlying roseness. Children learned how to speak like rats learned how to press levers. Words reflected sensory associations. Language had been solved.

B. F. Skinner, the psychologist who most sincerely applied the behaviorist work on animals to humans, even tried to explain literature in terms of stimuli and response. In 1934 Skinner wrote an essay in the *Atlantic Monthly* entitled "Has Gertrude Stein a Secret?" In the essay, he argued that Stein's experimental prose was really an experimental demonstration of behaviorism. According to Skinner, Stein was merely expressing the automatic verbal responses that we have to any specific stimulus. Her art was just involuntary reflex, the mutterings of an "unread and unlearned mind."

Stein objected fiercely to Skinner's critical theories. She believed that her writing was proof of the opposite psychology. "No, it is not so automatic as he [Skinner] thinks," she wrote in a letter to the editor of the *Atlantic*. "If there is anything secret it is the other way. I think I achieve by [e]xtra consciousness, [e]xcess, but then what is the use of telling him that . . ." While Skinner assumed that all of Stein's sentences were under "the control of a present sensory stimulus," most of Stein's descriptions celebrated a world too absurd to exist. In *Tender Buttons,* when she defines "Red Roses" as "a pink cut pink, a collapse and a sold hole, a little less hot," she knows that none of her words refers to anything factual. The color pink can't be cut and holes aren't sold. Buttons are never tender.

By flirting with absurdity, Stein forced us to acknowledge what Skinner ignored: the innate structure of language. While the behaviorists believed that grammar consisted of lots of little rules that we learned through parental nagging and teacher feedback, Stein realized that the mind wasn't so constrained. There was no grammatical rule that required us to call the rose red, or to not write "pink cut pink." After all, she made a living out of combining words in all sorts of unprecedented ways. No one had taught her how to write. The shock of the new — Stein's inexhaustible ability to invent original and ridiculous sentences — was what the behaviorists couldn't explain.

But where does this newness come from? How does the structure of language generate such a limitless array of possible expressions? Stein's startling insight was that our linguistic structures are *abstract*. Although Skinner argued that our grammar mandated specific words in specific contexts, Stein knew that he was wrong. In her writing, she shows us that our grammar only calls for the use of certain *kinds* of words in certain *kinds* of contexts. In *Tender Buttons*, when Stein describes "A Red Hat" as "A dark grey, a very dark grey," she is demonstrating this cognitive instinct. Although we rarely talk about red being gray, there is no linguistic rule prohibiting such a sentence. As long as the nouns, verbs, and adjectives are arranged in the right syntactical order, then any noun, verb, or adjective will do. We can describe the red hat any way we want. The same syntax can support an infinite number of sentences, even though many of those sentences would be meaningless.

This is an extremely weird way of imagining language. After all, the purpose of language is communication. Why, then, would its structure (syntax) operate independently of its meaning (semantics)? Aren't structure and function supposed to be intertwined? But Stein's dadaistic art declared that this wasn't the case. She defined language not in terms of its expressive content — her writing rarely makes sense — but in terms of its hidden structure. When meaning was stripped away, that is what remained.

Noam Chomsky

On September 11, 1956, at a meeting at MIT of the Institute for Radio Engineers, three new ideas entered the scientific canon. Each of these ideas would create a new field. All three would irrevocably alter the way we think about how we think.

The first idea was presented by Allen Newell and Herbert Simon. In their brief talk, they announced the invention of a machine capable of solving difficult logical problems. Essentially, their programming code translated the language of philosophical logic into computer-speak, finding engineering equivalents for logical tricks such as syllogisms and if-then statements. In fact, their machine was so effective that it solved thirty-eight of the first fifty-two proofs of A. N. Whitehead's and Bertrand Russell's *Principia Mathematica*. It even found a more elegant proof for one of Russell's problems. Intelligence had become "artificial." The mind had been faked.

The psychologist George Miller presented the second idea that day, wittily summarized by the title "The Magical Number Seven, Plus or Minus Two." Miller's idea was simple: the mind has limits. Our short-term memory, Miller said, can only contain about seven random bits before forgetfulness begins to intrude.* This is why all the random labels in our life, from phone numbers to license plates to Social Security numbers, are limited to seven digits (plus or minus two).

But Miller didn't stop there, for he knew that the mind didn't really deal in bits. We are constantly re-coding our sensations, discovering patterns in the randomness. This is how we see reality: not as bits, but as chunks. As Miller observed at the end of his paper, almost as an afterthought, "the traditional experimental psychologist has contributed little or nothing to the analysis [of re-coding]." Science had ignored the way the mind actually works, the way we make sense of reality by chunking its various parts together. From Miller's casual observation, cognitive psychology was born.

The last idea presented to the lucky radio engineers that day was by Noam Chomsky, a twenty-seven-year-old linguist with a penchant for big ideas. Chomsky's paper was entitled "Three Models for the Description of Language," but it was really about how one model — the finite-state approach to language, which grew out of behaviorism — was absurdly wrong. This linguistic theory tried to

* William James, as usual, was there first. Using a handful of marbles, James had shown more than sixty years earlier that the mind could process only limited amounts of information. His experiment was simple: he picked up an unknown number of marbles and tossed them into a box. While the marbles were in the air, he tried to guess how many he had thrown. He discovered that he could actually see the number of marbles he had tossed — without consciously counting them — as long as he hadn't tossed more than five or six.

reduce grammar to the laws of combinatorial statistics, in which each word in a sentence is generated from the previous word. A noun causes a verb, which causes another noun. (Add adjectives to taste.) The specific word choice is dictated by the laws of probability. Thus, "Roses are red" is a more likely phrase than "Pink cut pink." (According to this theory, Gertrude Stein had penned some of the most improbable sentences in history.)

In his technical lecture, Chomsky explained why language is not merely a list of words statistically strung together. His argument revolved around two separate examples, both of which sounded a lot like Stein.

Chomsky's first example was the sentence "Colorless green ideas sleep furiously." According to the statistical model of language, this sentence was technically impossible. No finite-state device could ever create it. Ideas don't sleep, and the probability of *colorless* being followed by the color *green* was exactly zero. Nevertheless, Chomsky demonstrated that the ridiculous sentence was grammatically feasible. Like Stein in *Tender Buttons,* Chomsky was using some suspiciously significant nonsense to prove that the structures governing words existed independently of the words themselves. Those structures came from the mind.

Chomsky's second argument was even more devastating. He realized that any linguistics dependent on statistics — as opposed to innate grammatical structures — was marred by one fatal flaw: it had no memory. Because a finite-state device created a sentence by adding one word at a time, it remembered only the previous word in the sentence and forgot about everything that came before. But Chomsky's insight was that certain sentences contain long-distance dependencies, in which the placement of any single word is derived from words much earlier in the sentence. Chomsky used either-or and if-then statements to prove his point. For example, the sentence "If language is real, then the finite-state model is false" contains a grammatical structure that can't be computed word by word, from left to right. Whenever we see an *if* we know to expect a *then,* but not right away. Unfortunately for the finite-state approach, by the time the statistical machine has gotten to the *then* it has forgotten all about the *if*. Although his argument was dense with linguistic lingo, Chomsky made his moral clear: "There are processes of language formation that this elementary model of language [the finite-state model] is intrinsically incapable of handling." This is because not every sentence is simply the sum of its separate words. In Chomsky's linguistics, our words are surrounded by syntactical interconnections, what he would later call the "deep structure of language."

The art of Gertrude Stein overflows with sentences that anticipate Chomsky's argument. For example, many of Stein's tedious repetitions — "I love repeating monotony," she said — actually reveal the deep structure of our sentences, the way our words are all entangled with one another. In *Tender Buttons*, as Stein struggles to define *vegetable*, she wonders: "What is cut. What is cut by it. What is cut by it in." The point of these sentences is that the same list of words can mean many different things. In the question "What is cut," the *what* refers to the thing that has been cut. However, when Stein wonders, "What is cut by it in," she is making *what* refer to *where* the cut thing is. She is demonstrating that a distant word can modify an earlier word, just as *in* modifies *what*. Of course, a finite-state device can't understand this sentence, since it computes meaning in a single direction only, from left to right. Like Chomsky's if-then phrases, Stein's sentence depends entirely upon its hidden syntax, upon the long-distance dependencies that bind its separate words together. Only a real mind can read Stein.

The Chomskian paradigm shift started that day, but it was only just beginning. Paradigm shifts take time, especially when they are written in the tortuous jargon of linguistics. Chomsky knew that before he could become right, he would have to prove everyone else wrong. His next target was the same figure Gertrude Stein had criticized more than twenty years earlier: the behaviorist B. F. Skinner. In 1959, two years after his epic but technical *Syntactic Structures* came out, Chomsky turned his review of Skinner's *Verbal Behavior* into a thirty-two-page manifesto. Chomsky's analysis was clearly and boldly written, and it marked the start of his career as a public intellectual.

In the review, Chomsky pointed out the dark flaw in the behaviorist explanation of words and rules. Language, Chomsky said, is *infinite*. We are able to create new sentences of any possible length, expressions never before imagined by another brain. This boundless creativity — best evidenced by the nonconformity of Stein's prose — is what separates human language from all other animal forms of communication. And unless we have an infinite set of reflexes, behaviorism cannot explain the infinite number of possible sentences.

So what is the proper way to conceive of language? Like Stein, Chomsky insisted that linguistics focus on the *structure* of language and not simply its individual words and phonetic tics. While linguists before Chomsky were content with classification and observation — they saw themselves as verbal botanists — Chomsky demonstrated that all their data missed the point. To see what Chomsky wanted us to see, we had to zoom out. Viewed from the lofty structuralist perspective, it suddenly became clear that every language — from English to Cantonese — was actually the same. While the words might be different, they shared the same subterranean form. Therefore, Chomsky hypothesized the existence of a universal grammar built into the brain. (As the New Testament preaches, "And the Word became flesh.") It is this innate language apparatus that lets us order words, composing them within a structure that is at once subtle and inescapable.

And while some of the details of Chomskian linguistics remain controversial, it is now clear that the deep structure of language is really an a priori instinct. The best evidence for this universal grammar comes from studies of the deaf in Nicaragua. Until the early 1980s, the deaf citizens of Nicaragua remained tragically isolated. The country didn't have a sign language, and deaf children were confined to overcrowded orphanages. However, when the first school for the deaf was founded, in 1981, the situation immediately began to improve. The children were never taught sign language (there were no teachers), but they suddenly began to speak with their hands. A makeshift vocabulary spontaneously evolved.

But the real transformation occurred when younger deaf students were introduced to this newly invented sign language. While older students were forced to converse in relatively imprecise terms, these second-generation speakers began to give their language a structure. No one had taught them grammar, but they didn't have to be taught. Just as Chomsky's theory had predicted, the young children imposed their innate knowledge onto their growing vocabulary. Verbs became inflected. Adjectives became distinct from nouns. Concepts that older speakers conveyed with single signs were now represented by multiple signs enclosed within a sentence. Although these Nicaraguan children had never known language, they invented their own. Its grammar looks a lot like every other human grammar. Stein was right: "There is only one language."*

Once Stein realized — fifty years before Chomsky — that the structure of language was unavoidable, she set out to make that structure palpably obvious. As Stein observed in a pithy anticipation of Chomskian linguistics, "Everybody said the same thing over and over again with infinite variations but over and over again." What Stein wanted to do was see the source of this sameness, to cut words until their structure showed through.

Of course, the presence of linguistic structures is hard to reveal. They are designed to be invisible, the clandestine scaffolding of our sentences. Stein's insight was that the reader was only aware of grammar when it was *subverted*. Just as Stravinsky had exposed the conventions of music by abandoning the conventions of music, so

---

* Another powerful example of the innateness of language comes from studies of slave and servant plantations in which a polyglot mixture of adult laborers evolved a common communicative system to help them talk to one another. At first, these laborers spoke in pidgin, which is a rudimentary language with little in the way of grammar. However, the children born in these plantations quickly transcended the limits of pidgin and developed various creole languages. These languages differed substantially from the pidgins in that they had all the grammatical features of established languages.

Stein demonstrated the power of grammar by abandoning grammar. In her art, she often tried to see how far she could push the envelope of bad syntax, morphology, and semantics. What would happen if she structured her writing based on the sound of her words, rather than their definitions? Could she write an entire novel only "in enormously long sentences that would be as long as the longest paragraph"? Or how about a sentence without any punctuation at all? Why do we write like this, and not like that?

What Stein discovered was a writing style that celebrated its grammatical mistakes. In her most radical prose, she manages to make us conscious of all the linguistic work that is normally done unconsciously. We notice the way verbs instantly get conjugated (even irregular verbs), the way nouns naturally become plural, and the way we amend articles to fit their subjects. Stein always said that the only way to read her writing was to proofread it, to pay acute attention to all the rules she violates. Her errors trace the syntactical structures we can't see, as our "inside becomes outside."* Stein showed us what we put into language by leaving it out.

The Meaning of Meaninglessness

The problem with difficult prose is its difficulty. T. S. Eliot may have made difficulty cool when he said that poets should be difficult, but he probably wasn't thinking of Stein's *The Making of Americans*, which has more than a thousand pages of repetitive nonnarrative. Before Stein's sentences can be understood (let alone enjoyed), they require a stubborn persistence on the part of the reader. They de-

---

* This isn't as strange a method as it might seem. Ludwig Wittgenstein hit upon a similar method for his philosophy, which, like Stein's writing, was interested in the uses of language to the exclusion of almost everything else. Wittgenstein once said that he worked by "mak[ing] a tracing of the physiognomy of every [philosophical] error." Only by mapping out mistakes could he see how best to proceed. Samuel Beckett also subscribed to Stein's literary approach. "Let us hope that a time will come," Beckett wrote, "when language is most efficiently used where it is being misused. To bore one hole after another in it, until what lurks behind it — be it something or nothing — begins to seep through; I cannot imagine a higher goal for a writer today."

mand time and more time, and even then whole paragraphs insist on remaining inscrutable. Although Stein is often funny, she is rarely fun. Sometimes, her confidence in her genius feels like insolence.

Nevertheless, Stein's fractal prose sets the stage for a less difficult avant-garde. "When you make a thing," Stein confessed in *The Autobiography*, "it is so complicated making it that it is bound to be ugly, but those that do it after you they don't have to worry about making it and they can make it pretty, and so everybody can like it when the others make it."

Literary history has borne out Stein's aesthetic theory. Without her original ugliness, it is hard to imagine the terse prettiness of Ernest Hemingway. When he was a reporter in Paris, Stein told Hemingway to quit his job and begin his novel. "If you keep on doing newspaper work," she said, "you will never see things, you will only see words." Stein later bragged that Hemingway learned how to write by correcting her own rough drafts. Hemingway's stripped-down sentences, drastically short and unadorned (except when they were long and ungrammatical), echo Stein's more stringent experiments. As Hemingway once joked to Sherwood Anderson, "Gertrude Stein and me are just like brothers."

Stein's own literary legacy has been shaped by her difficulty. While Hemingway's novels have inspired Hollywood movies and endless imitators, Stein's art is kept afloat by academics. If she is remembered today outside college campuses and histories of cubism, it is for a single cliché, one that is almost impossible to forget: "A rose is a rose is a rose is a rose."* Although Stein used this aphorism as a decoration for dinner plates, it now represents everything she wrote. This is the danger of avoiding plots.

Stein would be disappointed by her lack of influence. A woman of immense ambition, she hoped that her literature would save the

---

* "The roses under my window make no reference to former roses or better ones; they are what they are.... There is simply the rose," as Ralph Waldo Emerson said. He probably got the idea from Shakespeare, who has Juliet ponder the meaning of words: "What's in a name? That which we call a rose / By any other word would smell as sweet."

English language. "Words had lost their value in the nineteenth century," Stein lamented, "they had lost much of their variety and I felt that I could not go on that I had to recapture the value of the individual word." Her plan was simple. First, she would show us that words have no inherent meaning. For example, when she wrote, "A rose is a rose is a rose," what she was really trying to demonstrate was that a "rose" is *not* a rose. By repeating the noun again and again, Stein wanted to separate the signifier from what it signified, to remind us that every word is just a syllable of arbitrary noise. (As Tennyson observed, we can make our own name seem strange simply by repeating it a few times fast.) According to Stein's scheme, this act of deconstruction would allow us to reconstruct our language, to write without lapsing into cliché. She used her "rose" sentence as an example of how such a process might work: "Now listen! I'm no fool," she told Thornton Wilder in 1936. "I know that in daily life we don't go around saying 'is a . . . is a . . . is a . . .' But I think that in that line the rose is red for the first time in English poetry for a hundred years."

But Stein's grand plan ran into a serious problem. No matter how hard she tried — and Stein tried very hard — her words refused to become meaningless. The rose never surrendered its stale connotations. *Tender Buttons* didn't erase the definitions of *tender* or *buttons*. The words of English easily survived Stein's modernist onslaught. After a few years, her revolution petered out, and writers went back to old-fashioned storytelling. (It didn't help, of course, that her books rarely circulated beyond the Left Bank.)

Why couldn't Stein reinvent the dictionary? Why was saying nothing so damn hard? The answer returns us to her earlier discovery: the structure of language. Because words are always interconnected by syntax, they can never say nothing. Meaning is contextual and holistic, and no word exists alone. This is why Stein's silliest sentences continue to inspire all sorts of serious interpretations. (As William James observed in his *Principles of Psychology,* "Any collocation of words may make sense — even the wildest words in a dream — if one only does not doubt their belonging together.") In a

1946 interview, given just a few months before her death, Stein finally admitted defeat. She would never save language by dismantling it, because language couldn't be dismantled. "I found out that there is no such thing as putting them [words] together without sense. It is impossible to put them together without sense. I made innumerable efforts to make words write without sense and found it impossible. Any human being putting down words had to make sense out of them."

Ironically, Stein's experimental failure, her inability to make her prose entirely meaningless, was her greatest achievement. Although she aimed for obscurity, her art still resonates. Why? Because the structure of language — a structure that her words expose — is part of the structure of the brain. No matter how abstract Stein made her writing, she still wrote from inside our language game, constrained by an instinct as deep as it is universal. The innate grammar that Chomsky would later discover was the one instinct that not even Stein could write without. "How can grammar be?" she asked herself in *How to Write*. "Nevertheless" was her answer.

Chapter 8

Virginia Woolf

The Emergent Self

> Psychology which explains everything
> explains nothing.
>
> — Marianne Moore

IN 1920, AFTER WRITING two novels with a conventional Victorian narrator (the kind that, like an omniscient God, views everything from above), Virginia Woolf announced in her diary: "I have finally arrived at some idea of a new form for a new novel." Her new form would follow the flow of our consciousness, tracing the "flight of the mind" as it unfolded in time. "Only thoughts and feelings," Woolf wrote to Katherine Mansfield, "no cups and tables."

This modernist style was a radical shift in perspective. The eminent novelists of her time — "Mr. Wells, Mr. Bennett, and Mr. Galsworthy" — ignored the mind's interiors. "They have looked," Woolf wrote, "at factories, at Utopias, even at the decoration and upholstery of the carriage; but never at life, *never at human nature.*" Woolf wanted to invert this hierarchy. "Examine for a moment an ordinary mind on an ordinary day," she wrote in her essay "Modern Fiction." "Is it not the task of the novelist to convey this varying, this unknown and uncircumscribed spirit, whatever aberration or complexity it may display, with as little mixture of the alien and external as possible?"

But the mind is not an easy thing to express. When Woolf looked inside herself, what she found was a consciousness that never stood still. Her thoughts flowed in a turbulent current, and every moment ushered in a new wave of sensation. Unlike the "old-fashioned novelists," who treated the mind as a static thing, Woolf described the mind as neither solid nor certain. Instead, it "was very erratic, very undependable — now to be found in a dusty road, now in a scrap of newspaper in the street, now in a daffodil in the sun." At any given moment, she seemed to be scattered in a million little pieces. Her brain was barely bound together.

And yet, it *was* bound together. Her mind was made of fragments, but it never came undone. She knew that something kept us from disintegrating, at least most of the time. "I press to my centre," Woolf wrote in her diary, "and there is something there."

Woolf's art searched for whatever held us together. What she found was the self, "the essential thing." Although the brain is just a loom of electric neurons, Woolf realized that the self makes us whole. It is the fragile source of our identity, the author of our consciousness. If the self didn't exist, then we wouldn't exist. "One must have a whole in one's mind," Woolf said, "fragments are unendurable."

But if the mind is so evanescent, then how does the self arise? Why do we feel like more than just a collection of disconnected thoughts? Woolf's revelation was that we emerge from our own fleeting interpretations of the world. Whenever we sense something, we naturally invent a subject for our sensation, a perceiver for our perception. The self is simply this subject; it is the story we tell ourselves about our experiences. As Woolf wrote in her unfinished memoir, "We are the words; we are the music; we are the thing itself."

At the time, this was a surreal idea. Scientists were busy embracing the power of materialism. Anatomy promised to explain everything. (As James Joyce noted, "The modern spirit is vivisective.") The self was just another trick of matter, which time and experiments would discover. But Woolf knew that the self was too profound to be found. In her modernist novels, she wanted to expose

Portrait photograph of Virginia Woolf,
by George Charles Beresford, 1902

our ineffability, to show us that we are "like a butterfly's wing . . . clamped together with bolts of iron." If the mind is a machine, then the self is its ghost. It is what cannot be seen.

Almost a century later, the self remains elusive. Neuroscience has ransacked the brain and dissected the cortex, but it has not found our source. Although experiments have confirmed many of Woolf's startling insights — the mind is made of fragments, and yet these fragments are bound into being — our mystery persists. If we want to understand ourselves, Woolf's art is our most revealing answer.

The Split Minds of Modernism

Woolf's writing style was deeply rooted in her own experience of the brain: She was mentally ill. All her life, she suffered from periodic nervous breakdowns, those horrible moments when her depression became suffocating. As a result, Woolf lived in fear of her own mind, exquisitely sensitive to its fevered "vibrations." Introspection was her only medicine. "My own psychology interests me," she confessed to her journal. "I intend to keep full notes of my ups and downs for my private information. And thus objectified, the pain and shame become at once much less." When all else failed, she used her sardonic humor to blunt the pain: "I feel my brains, like a pear, to see if it's ripe; it will be exquisite by September." And while she complained to E. M. Forster and others about her doctors and their syrups, about the pain and torpor of being forced to lie in bed, she also acknowledged the strange utility of her illness. Her incurable madness — this "whirring of wings in the brain" — was, in some ways, strangely transcendental: "Not that I haven't picked up something from insanities and all the rest. Indeed, I suspect they've done instead of religion."*

Woolf never recovered. Her constant state of reflection, her wariness of hints of the return of her devastating depression, scarred her writing. *Nerves* was one of her favorite words. Its medical variations — neurosis, neurasthenia, nervous breakdown, neurasthenic — continually entered her prose, their sharp, scientific pang contradicting the suppleness of her characters' internal soliloquies. In Woolf's diary, notes on form were always interwoven with comments on headaches.

But her illness also gave her experimental fictions a purpose, a way "of depositing experience in a shape that fitted it." After each depressive episode, she typically experienced a burst of creativity as she filled her journal with fresh insights into the workings of her

---

* This is not to aestheticize sickness. Insanity is real: on March 28, 1941, Woolf put stones in her coat pockets, walked to the river, and drowned herself.

own "difficult nervous system." Forced by her doctors to lie in bed, she passed the time by staring at the ceiling and contemplating her own brain. She decided that she had "no single state." "It's odd how being ill," she observed, "splits one up into several different people." At any given moment, she was both mad and lucid, ingenious and insane.

What Woolf learned about the mind from her illness — its quicksilverness, its plurality, its "queer conglomeration of incongruous things" — she transformed into a literary technique. Her novels are about the difficulty of knowing people, of saying that "they were this or were that." "It is no use trying to sum people up," she writes in *Jacob's Room*. Although the self seems certain, Woolf's writing exposes the fact that we are actually composed of ever-changing impressions that are held together by the thin veneer of identity. Like Septimus, the prophetic madman whose suicide is the climax of *Mrs. Dalloway*, we live in danger of coming apart. The mystery of why we do not *always* come apart is the animating tension in her art.

"At [the age of] forty," Woolf wrote in her journal in 1922, "I am beginning to learn the mechanism of my own brain." That same year, Woolf began writing *Mrs. Dalloway*, her literary response to *Ulysses*. Like Joyce, she set her novel on a weekday in a bustling city. Her main character, Clarissa Dalloway, is neither heroic nor tragic, but simply one of "those infinitely obscure lives that must be recorded." As Woolf liked to remind herself, "Let us not take it for granted that life exists more in what is commonly thought big than in what is commonly thought small."

The novel famously begins with Mrs. Dalloway going to buy the flowers herself. The June day will consist mostly of this sort of errand, but Woolf, as always, manages to expose the profound in the quotidian. This is life as it's lived, she says: our epiphanies inseparable from our chores, our poetry intermingled with the prose of ordinary existence. A single day, rendered intensely, can become a vivid window into our psychology.

Woolf uses *Mrs. Dalloway* to demonstrate the mind's fragility. She interweaves Clarissa's party with the suicide of Septimus Smith, a veteran of the Great War who has become a shell-shocked poet.* Dr. Bradshaw, Septimus's "obscurely evil" doctor, tries to cure his madness by imposing a regimen of "proportion," but the medicine just makes everything worse. Bradshaw's insistence that Septimus's illness was "physical, purely physical" causes the poet to kill himself. His self has fallen apart, and it cannot be put back together.

At her party, Clarissa hears of Septimus's suicide, and she is devastated. Although she had never met Septimus, Clarissa "felt very much like him." Like Septimus, she knows that her own self is frighteningly precarious, lacking "something central which permeated." When Dr. Bradshaw shows up at her party, she suspects him of committing "some indescribable outrage — forcing your soul, that was it."

But that is where the parallels end. Unlike Septimus, Clarissa compensates for her fragmentary being. Although she doesn't believe in an immortal soul — Clarissa is a skeptical atheist — she has developed a "transcendental theory" about "the unseen part of us." Doctors like Bradshaw deny the self, but Clarissa doesn't; she knows that her mind contains an "invisible center." As the novel unfolds, this center begins to emerge: "That was her self," thinks Clarissa, as she stares into the mirror. "Pointed; dartlike; definite. That was her self when some effort, some call on her to be her self, drew the parts together, she alone knew how different, how incompatible and composed so for the world only into one centre, one diamond, one woman who sat in her drawing room . . ." The point is that Mrs. Dalloway *does* draw herself together. She makes herself real, creat-

---

* As Woolf began to write the novel, in the fall of 1922, shell shock was starting to be recognized as a genuine psychiatric illness. Elaine Showalter has pointed out that doctors treated this new scourge using the same blunt tools they had been using on women such as Woolf for more than twenty years. These treatments included drugging the patients with bromides, confining them to bed and force-feeding them milk, and pulling their teeth, which was believed to lower the temperature of the body. Other unfortunate patients got the fever cure, in which psychosis was treated with an injection of malaria, tuberculosis, or typhoid. The Nobel Prize was awarded for this sadistic treatment in 1927.

Virginia Woolf's notebook for Mrs. Dalloway, *1925*

ing a "world of her own wherever she happens to be." This is what we all do every day. We take our scattered thoughts and inconstant sensations and we bind them into something solid. The self invents itself. The last line of the novel affirms Mrs. Dalloway's tenuous presence: "For there she was."

To the Lighthouse, Woolf's next novel, ventured even deeper into the turbulent mind. Writing the book, Woolf said, was the closest she ever came to undergoing psychoanalysis. After a long summer of illness, the prose just poured out of her, like a confession. The novel itself has little plot. There is almost a trip to the lighthouse, and then it's dinnertime, and then time passes, and then there *is* a trip to the lighthouse, and then Lily, a painter, finishes her painting. But despite the paucity of events, the novel feels frantic, dense with the process of minds flowing through time. The narrative is constantly being interrupted by thought, by thought about thought, by thought about reality. A fact is stated by someone (in thought or aloud), and then it is contradicted. Often, the same brain contradicts itself.

According to Woolf, this mental disorder is an accurate description of our mental reality. The self emerges from the chaos of consciousness, a "kind of whole made of shivering fragments." In her essay "Modern Fiction" — one of her most potent descriptions of modernist aspirations — Woolf defined her new literary style in psychological terms. "The mind receives a myriad of impressions," Woolf wrote. "From all sides they come, an incessant shower of innumerable atoms; and as they fall, they shape themselves into the life of Monday or Tuesday . . . Let us [the modern novelist] record the atoms as they fall upon the mind in the order in which they fall, let us trace the pattern, however disconnected and incoherent in appearance, which each sight or incident scores upon the consciousness."

To the Lighthouse is full of such falling thoughts. The characters overflow with impermanent impressions and inchoate feelings. This is perhaps most true of Mrs. Ramsay, the mother at the center of the novel. As she thinks of her husband, a philosopher writing an encyclopedia of philosophy that is stuck on the letter *Q*, her mind is torn in different directions. Mr. Ramsay has just denied their son James a trip to the lighthouse due "to the barometer falling and the wind due west." Mrs. Ramsay thinks her husband is being unfair: "To pursue truth with such astonishing lack of consideration for other people's feelings, to rend this thin veil of civilization so wantonly, so brutally, was to her so horrible an outrage of human decency." But then, one sentence later, Mrs. Ramsay experiences a total reversal in her thoughts: "There was nobody she reverenced as she reverenced him . . . She was nothing but a sponge sopped full of human emotions. She was not good enough to be his shoe strings."

For Woolf, Mrs. Ramsay's incoherent feelings are examples of reality honestly transcribed. By taking us inside the frayed minds of her characters, she reveals our own fragility. The self is no single thing and the stream of our consciousness just flows. At any given moment, we are at the whim of feelings we don't understand and sensations we can't control. While Mr. Ramsay believes that "thought is like the keyboard of a piano . . . all in order," Mrs. Ramsay knows that the mind is always "merging and flowing and creating." Like the Hebrides weather, change is its only constant.

Woolf's writing never lets us forget the precariousness of our being. "What does one mean by 'the unity of the mind,'" Woolf wondered in *A Room of One's Own*, "it [the mind] seems to have no single state of being." She wanted her readers to become aware of "the severances and oppositions in the mind," the way consciousness can "suddenly split off." At the very least, Woolf writes, one must always recognize "the infinite oddity of the human position." Although the self seems everlasting — "as solid as forever" — it lasts only for a moment. We pass "like a cloud on the waves."

This vision of a mercurial mind, a self divided against itself, was one of the central tenets of modernism. Nietzsche said it first: "My hypothesis is the subject as multiplicity," he wrote in a terse summary of his philosophy. "I is another," Rimbaud wrote soon after. William James devoted a large portion of his chapter on the self in *The Principles* to "mutations of the self," those moments when we become aware of our "other simultaneously existing consciousnesses." Freud agreed, and he shattered the mind into a network of conflicting drives. T. S. Eliot converted this idea into a theory of literature, disowning "the metaphysical theory of the substantial unity of the soul." He believed that the modern poet had to give up the idea of expressing the "unified soul" simply because we didn't have one. "The poet has, not a 'personality' to express," Eliot wrote, "but a particular medium, which is only a medium and not a personality." Like so many modernists, Eliot wanted to pierce our illusions, revealing us not as we want to be, but as we are: just the rubble of being, some random scraps of sensation. Woolf echoed Eliot, writing in her diary that we are "splinters and mosaics; not, as they used to hold, immaculate, monolithic, consistent wholes."

Surreal as it seems, the modernists got the brain right. Experiment after experiment has shown that any given experience can endure for about ten seconds in short-term memory. After that, the brain exhausts its capacity for the present tense, and its consciousness must begin anew, with a new stream. As the modernists anticipated, the permanent-seeming self is actually an endless procession of disjointed moments.

Even more disconcerting is the absence of any single location in the brain — a Cartesian theater, so to speak — where these severed moments get reconciled. (What Gertrude Stein said about Oakland is also true of the cortex: "There is no there there.") Instead, the head holds a raucous parliament of cells that endlessly debate what sensations and feelings should *become* conscious. These neurons are distributed all across the brain, and their firing unfolds over time. This means that the mind is not a place: it is a process. As the influential philosopher Daniel Dennett wrote, our mind is made up "of multiple channels in which specialist circuits try, in parallel pandemoniums, to do their various things, creating Multiple Drafts as they go." What we call reality is merely the final draft. (Of course, the very next moment requires a whole new manuscript.)

The most direct evidence of our diffuseness comes from the shape of the brain itself. Though the brain is enclosed by a single skull, it is actually made of two separate lumps (the left and right hemispheres), which are designed to disagree with each other. Lily, the heroic painter of *To the Lighthouse*, got her anatomy exactly right: "Such was the complexity of things . . . to feel violently two opposite things at the same time; that's what you feel, was one; that's what I feel, was the other, and then they fought together in her mind, as now." As Lily notes, every brain is crowded with at least two different minds.

When the neuroscientists Roger Sperry and Michael Gazzaniga

first stated this idea in 1962, it was greeted with derision and skepticism.* Studies of patients with brain injuries had concluded that the left side of the brain was the conscious side. It was the seat of our soul, the place where everything came together. The other half of the brain — the right hemisphere — was thought to be a mere accessory. Sperry, in his 1981 Nobel lecture, summarized the prevailing view of the right hemisphere when he began studying it: The right hemisphere "was not only mute and agraphic but also dyslexic, word-deaf and apraxic, and lacking generally in higher cognitive function."

Sperry and Gazzaniga disproved this doctrine by investigating human split-brain patients with severed corpora callosa (the corpus callosum is the thin bridge of nervous tissue that connects the brain's left and right hemispheres). Neurologists had studied these patients before and found them essentially normal. (As a result of that finding, surgically dividing the brain became a common treatment for severe epilepsy.) This confirmed the neurologists' suspicion that consciousness only required the left hemisphere.

But Sperry and Gazzaniga decided to look a little closer. The first thing they did when studying split-brain patients was test the abilities of the right hemisphere when it was isolated. To their surprise, the right hemisphere wasn't silent or stupid; instead, it seemed to play an essential role in "abstraction, generalization, and mental association." Contrary to the dogma of the time, one of the brain's halves did *not* dominate and intimidate the other. In fact, these patients proved that the opposite was true: each lobe had a unique self, a distinct being with its own desires, talents, and sensations. As Sperry wrote, "Everything that we have seen so far indicates that the surgery has left these people with two separate minds, i.e., with two separate spheres of consciousness."

---

* It shouldn't have been. In 1908, the German neurologist Kurt Goldstein described a patient suffering from multiple strokes. A later autopsy revealed that she also suffered from a lesion in her corpus callosum. Goldstein observed the strangeness of her behavior: "On one occasion the hand grabbed her own neck and tried to throttle her, and could only be pulled off by force. Similarly, it tore off the bed covers against the patient's will . . ."

And while the corpus callosum lets each of us believe in his or her singularity, every *I* is really plural. Split-brain patients are living proof of our many different minds. When the corpus callosum is cut, the multiple selves are suddenly free to be themselves. The brain stops suppressing its internal inconsistencies. One patient reading a book with his left hemisphere found that his right hemisphere, being illiterate, was extremely bored by the letters on the page. The right hemisphere commanded the left hand to throw the book away. Another patient put on his clothes with his left hand while his right hand busily took them off. A different patient had a left hand that was rude to his wife. Only his right hand (and left brain) was in love.

But why are we normally unaware of this cortical conflict? Why does the self feel whole when it is really broken? To answer this question, Sperry and Gazzaniga mischievously flashed different sets of pictures to the right and left eyes of their split-brain patients. For example, they would flash a picture of a chicken claw to a patient's right eye and a picture of a snowy driveway to the left eye. The patient was then shown a variety of images and asked to pick out the image that was most closely associated with what he had seen. In a tragicomic display of indecisiveness, the split-brain patient's two hands pointed to two different objects. The right hand pointed to a chicken (this matched the chicken claw that the left hemisphere had witnessed), while the left hand pointed to a shovel (the right hemisphere wanted to shovel the snow). When the scientists asked the patient to explain his contradictory responses, he immediately generated a plausible story. "Oh, that's easy," he said. "The chicken claw goes with the chicken, and you need a shovel to clean out the chicken shed." Instead of admitting that his brain was hopelessly confused, he wove his confusion into a neat narrative.

Sperry and Gazzaniga's discovery of the divided mind, and the way we instinctively explain away our divisions, had a profound impact on neuroscience. For the first time, science had to confront the idea that consciousness emerged from the murmurings of the *whole* brain and not from just one of its innumerable parts. According to

Sperry, our feeling of unity was a "mental confabulation"; we invented the self in order to ignore our innate contradictions. As Woolf wondered in her essay "Street Haunting," "Am I here, or am I there? Or is the true self neither this nor that but something so varied and wandering that it is only when we give rein to its wishes and let it take its way unimpeded that we are indeed ourselves?"

Emergence

The poignant irony underlying Woolf's fiction is that although she set out to deconstruct the self, to prove that we were nothing but a fleeting "wedge of darkness," she actually discovered the self's stubborn reality. In fact, the more she investigated experience, the more necessary the self became to her. If we know nothing else, it is that we are here, experiencing this. Time passes and sensations come and go. But we remain.

Woolf's characters reflect her fragile faith in the self. In her novels, everything is seen through the subjective prism of an individual. Mr. Ramsay is different than Mrs. Ramsay. While he looks out at the clouded sky and thinks of rain, she wonders if the wind might change. Mrs. Dalloway, for all of her strange congruencies with Septimus, is *not* Septimus. She doesn't leap out of a window. She throws a party. No matter how modernist Woolf's prose became, the illusory self — that inexplicable essence that makes us ourselves, and not someone else — refused to disappear. "Did I not banish the soul when I began?" Woolf asked herself in her diary. "What happens is, as usual, life breaks in."

In her art, Woolf let life break in. She shows us our fleeting parts, but she also shows us how our parts come together. The secret, Woolf realized, was that the self *emerges* from its source. *Emerge* is the crucial word here. While her characters begin as a bundle of random sensations, echoing about the brain's electrical membranes, they instantly swell into something else. This is why, as Erich Auerbach pointed out in *Mimesis,* Woolf's modernist prose is neither a continuous transcription of the character's self-consciousness (as in Joyce's *Ulysses*) nor an objective description of events and perceptions (as in a typical nineteenth-century novel). Woolf's revelation was to merge these two polarities. This technique allows her to document consciousness as a *process,* showing us the full arc of our thought. The impersonal sensation is always ripening into a subjective experience, and that experience is always flowing into the next one. And yet, from this incessant change, the character emerges. Woolf wanted us to see both sides of our being, how we are "a thing that you could ruffle with your breath; and a thing you could not dislodge with a team of horses." In her fiction, the self is neither imposed nor disowned. Rather, it simply arises, a vision stolen from the flux.

But *how* does the self arise? How do we continually emerge from our sensations, from the "scraps, orts and fragments" of which the mind is made?

Woolf realized that the self emerges via the *act of attention.* We bind together our sensory parts by experiencing them from a particular point of view. During this process, some sensations are ignored, while others are highlighted. The outside world gets thoroughly interpreted. "With what magnificent vitality the atoms of my attention disperse," Woolf observed, "and create a richer, a stronger, a more complicated world in which I am called upon to act my part."

Woolf's finest description of the process of attention comes in the dinner scene of *To the Lighthouse,* after the *boeuf en daube* is served. Mrs. Ramsay, the content matriarch, drifts into a reverie, her mind settling in that "still space that lies at the center of things." She has stopped listening to the dinner conversation (they were only talking about "cubes and square roots") and has begun contemplating the bowl of fruit at the center of the table. With a "sudden exhilaration," her mind becomes "like a light stealing under water," piercing through the "the flowing, the fleeting, the spectral." Mrs. Ramsay is now paying attention: the tributaries of her sensation have flowed into the serial stream of consciousness.

Woolf's prose stretches this brief second of brain activity into an extended soliloquy, as her words intimately observe the flow of Mrs. Ramsay's mind. We watch her eyes drift over to the dish of fruit, and follow her gaze as it settles on the purple grapes and then the ripe yellow pear. What began as an unconscious urge — Mrs. Ramsay stares at the fruit "without knowing why" — is now a conscious thought. "No," Mrs. Ramsay thinks to herself, "she did not want a pear."

In this moment of attention, Mrs. Ramsay's mind has remade the world. Her self has imposed itself onto reality and created a conscious experience. "Beneath it is all dark," Mrs. Ramsay thinks. "But now and again we rise to the surface and that is what you see . . ." Now and again, attention binds together our parts, and the self transforms ephemeral sensations into a "moment of being." This is the process unfolding inside Mrs. Ramsay's mind. Everything is evanescent, and yet, for the reader, Mrs. Ramsay always seems real. She never wavers. We never doubt her existence, even when we see the impermanence of her source. "Of such moments," Mrs. Ramsay thinks, "the thing is made that endures."

But how do we endure? How does the self transcend the separateness of its attentive moments? How does a process become us? For Woolf, the answer was simple: *the self is an illusion*. This was her final view of the self. Although she began by trying to dismantle the stodgy nineteenth-century notion of consciousness, in which the self was treated like a "piece of furniture," she ended up realizing that the self actually existed, if only as a sleight of mind. Just as a novelist creates a narrative, a person creates a sense of being. The self is simply our work of art, a fiction created by the brain in order to make sense of its own disunity. In a world made of fragments, the self is our sole "theme, recurring, half remembered, half foreseen." If it didn't exist, then nothing would exist. We would be a brain full of characters, hopelessly searching for an author.

Modern neuroscience is now confirming the self Woolf believed in. We invent ourselves out of our own sensations. As Woolf anticipated, this process is controlled by the act of attention, which turns our sensory parts into a focused moment of consciousness. The fictional self — a nebulous entity nobody can find — is what binds these separate moments together.

Take the act of looking at a bowl of fruit. Whenever we pay attention to a specific stimulus — like a pear on a dinner table — we increase the sensitivity of our own neurons. These cells can now see what they would otherwise ignore. Sensations that were invisible suddenly become visible, as the lighthouse of attention selectively increases the firing rate of the neurons it illuminates. Once these neurons become excited, they bind themselves together into a temporary "coalition," which enters the stream of consciousness. What's important to note about this data is that attention seems to be operating in a top-down manner (what neuroscience calls "executive control"). The illusory self is causing very real changes in neuronal firing. It's as if the ghost is controlling the machine.

However, if the self does *not* pay attention, then the perception never becomes conscious. These neurons stop firing, and the sliver of reality they represent ceases to exist. When Mrs. Ramsay tunes out the dinner-table conversation and focuses instead on the fruit, she is literally altering her own cells. In fact, our consciousness seems to require such a discerning self: we only become aware of the sensation *after* it has been selected. As Woolf put it, the self is "our central oyster of perceptiveness."

The most startling evidence of the power of the conscious self comes from patients who are unable to pay attention. This insight came from an unlikely place: the blind. Before Lawrence Weiskrantz began his investigations in the early 1970s, science had assumed that lesions in the primary visual areas (the V1) caused irreparable blindness. They were wrong.

Lesions in the V1 only cause *conscious* blindness, a phenomenon Weiskrantz named "blindsight." Although these patients think they are blind, they can actually see, at least unconsciously. What they are missing is awareness. While their eyes continue to transmit visual information, and undamaged parts of their brains continue to process this information, blindsight patients are unable to consciously access what their brains know. As a result, all they see is darkness.

So how can you tell blindsight and blindness apart? Blindsight patients exhibit an astonishing talent. On various visual tasks they perform with an aptitude impossible for the totally blind. For example, a blindsight patient can "guess" with uncanny accuracy whether a square or a circle had just been shown, or whether a light had been flashed. While they have no explicit awareness of the light, they can still respond to it, albeit without knowing what they are responding to. Brain scans confirm their absurd claims, as the areas associated with self-awareness show little or no activity, while the areas associated with vision show relatively normal activity.

This is what makes blindsight patients so poignantly fascinating: their consciousness has been divorced from their sensations. Although the brain continues to "see," the mind can't pay attention to these visual inputs. They are unable to subjectively interpret the information entering the cortex. Blindsight patients are sad evidence that we have to transform our sensation — by way of the moment of attention, which is modulated by the self — before we can sense it.* A sensation separated from the self isn't a sensation at all.

Of course, the one thing neuroscience *cannot* find is the loom of cells that creates the self. If neuroscience knows anything, it is that there is no ghost in the machine: there is only the vibration of the machinery. Your head contains a hundred billion electrical cells, but not one of them is you or knows you or cares about you. In fact, you don't even exist. The brain is nothing but an infinite regress of matter, reducible to the callous laws of physics.

This is all undoubtedly true. And yet, if the mechanical mind is

---

* People with specific lesions on one side of the brain have confirmed this finding. These patients suffer from a syndrome known as visuospatial neglect. In severe cases, a victim of this syndrome is unable to pay attention to one half of the visual world. For example, a man will read only the pages on the left side of his book, or a woman will apply lipstick to only the right side of her lips. Nevertheless, patients normally remain able to respond to stimuli in their neglected receptive fields, even if they can't actually admit that such stimuli exist.

denied the illusion of a self, if the machine lacks a ghost, then every-thing falls apart. Sensations fail to cohere. Reality disappears. As Woolf wondered in *The Waves:* "How to describe the world seen without a self?" "There are no words," she answered, and she was right. Deprived of the fictional self, all is dark. We think we are blind.

Lily

The most mysterious thing about the human brain is that the more we know about it, the deeper our own mystery becomes. The self is no single thing, and yet it controls the singularity of our atten-tion. Our identity is the most intimate thing we experience, and yet it emerges from a shudder of cellular electricity. Furthermore, Woolf's original question — why the self feels real when it is not — remains completely unanswered. Our reality seems to depend upon a miracle.

In typically stubborn fashion, however, neuroscience has stam-peded straight into the mystery, attempting to redefine the incom-prehensible in terms of the testable. After all, the promise of mental reductionism is that no reference to higher-order functions (like ghosts or souls or selves) is required. Neurons, like atoms, explain everything, and awareness must percolate from the bottom up.

The most tractable scientific approach to the problem of con-sciousness is, not surprisingly, the search for its physical substrate. Neuroscience believes that if it looks hard enough it will be able to find the self's secret source, the fold of flesh that decides what to pay attention to. The technical term for this place is the "neural cor-relate of consciousness" (NCC).

Christof Koch, a neuroscientist at Caltech, is leading the search party. Koch defines the NCC as "the minimal set of neuronal events that gives rise to a specific aspect of a conscious percept." For exam-ple, when Mrs. Ramsay contemplates the bowl of fruit at the center of the table, her NCC (at least as Koch defines it) is the network of cells that create her consciousness of the pear. He believes that if science found the NCC, it could see exactly how the self emerges from its sensation. Our fountainhead would be revealed.

That sounds easy enough (science excels at uncovering material causality). But in actuality, the NCC is fiendishly elusive. Koch's first problem was finding an experimental moment where the unity of our consciousness is temporarily shattered, and thus vulnerable to reductionist inquiry. He settled on an optical illusion known as binocular rivalry to use as his main experimental paradigm. In theory, binocular rivalry is a simple phenomenon. We each have two eyeballs; as a result, we are constantly being confronted with two slightly separate views of the world. The brain, using a little unconscious trigonometry, slyly erases this discrepancy, fusing our multiple visions into a single image. (As split-brain patients demonstrate, we are built to ignore our own inconsistencies.)

But Koch decided to throw a wrench into this visual process. "What happens," he wondered, "if corresponding parts of your left and right eyes see two quite distinct images, something that can easily be arranged using mirrors and a partition in front of your nose?" Under ordinary circumstances, the brain superimposes the two separate images from our two separate eyes on top of each other. For example, if the left eye is shown horizontal stripes and the right eye is shown vertical stripes, a person will consciously perceive a plaid pattern. Sometimes, however, the self gets confused and decides to pay attention to the input of only one eye. After a few seconds, the self realizes the mistake, and it begins to pay attention to the *other* eye. As Koch notes, "The two percepts can alternate in this manner indefinitely."

The end result of all this experimentally induced confusion is that the subject becomes aware — if only for a moment — of the artifice underlying perception. He realizes that he has the input of two separate eyes, which see two separate things. Koch wants to know where in the brain the struggle for ocular dominance occurs. What neurons decide which eye to pay attention to? What cells impose a unity onto the sensory disarray? Koch believes that if he finds this place he will find one of our neural correlates of consciousness. He will discover where the self hides.

Despite its conceptual elegance, there are several serious problems with this experimental approach. The first problem is methodological. The brain is the universe's largest knot. Each of the brain's neurons is connected with up to a thousand other neurons. Consciousness derives its power from this recursive connectivity. After all, the self emerges not from some discrete Cartesian stage, but from the interactions of the brain *as a whole*. As Woolf wrote, "Life is not a series of gig lamps symmetrically arranged; life is a luminous halo . . . surrounding us from the beginning of consciousness to the end." Any reduction of consciousness to a single neural correlate — a "gig lamp" — is by definition an abstraction. While the NCC might describe where certain perceptual experiences take place, it will not reveal the origin of attention, or somehow solve the self, for those are emergent properties with no single source. The imponderable mystery Woolf wrote about will remain. Neuroscience must be realistic about what its experiments can explain.

The other big flaw in the NCC approach to consciousness is childishly simple and applies to *all* reductionist approaches to the mind. Self-consciousness, at least when felt from the inside, feels like more than the sum of its cells. Any explanation of our experience solely in terms of our neurons will never explain our experience, because we don't experience our neurons. This is what Woolf knew. She always believed that descriptions of the mind in purely physical terms — and the history of science is littered with such failed theories — were defined by their incompleteness. Such reductionist psychologies, Woolf wrote, "simplify rather than complicate, detract rather than enrich." They deny us our essential individuality, and turn "all our characters into cases." The mind, Woolf reminds us, cannot be solved by making every mind the same. To define consciousness solely in terms of oscillations in the prefrontal cortex is to miss out on our subjective reality. The self feels whole, but all science can see is its parts.

This is where art comes in. As Noam Chomsky said, "It is quite possible — overwhelmingly probable, one might guess — that we will always learn more about human life and personality from novels than from scientific psychology." If science breaks us apart, art puts us back together. In *To the Lighthouse*, Lily describes her artistic ambition: "For it was not knowledge but unity that she desired, not inscriptions on tablets, nothing that could be written in any language known to men, but intimacy itself, which is knowledge." Like Woolf, Lily wants to express our experience. She knows that this is all we can express. We are only intimate with ourselves.

The artist describes what the scientist can't. Though we are nothing but flickering chemicals and ephemeral voltages, the self seems real. In the face of this impossible paradox, Woolf believed that science must surrender its claims of absolute knowledge. Experience trumps the experiment. Since Woolf wrote her modernist novels, nothing has fundamentally changed. New psychologies have come and gone, but our self-awareness continues to haunt our science, a reality too real to be measured. As Woolf understood, the self is a fiction that cannot be treated like a fact. Besides, to understand ourselves as works of fiction is to understand ourselves as fully as we can. "The final belief," Wallace Stevens once wrote, "is to believe in a fiction, which you know to be a fiction, there being nothing else."

Although neuroscience is still impossibly far from a grand unified theory of consciousness, it has nevertheless confirmed the ideas in Woolf's art. Consciousness is a process, not a place. We emerge, somehow, from the moment of attention. Without the illusory self, we are completely blind.

But just because our essence is intangible doesn't mean that we should abandon all attempts to understand it. *To the Lighthouse*, a novel about the difficulty of knowing, ends with a discovery. In the momentous final scene, Woolf uses the character of Lily to show us how, despite our paradoxical source, we can nevertheless learn truths about ourselves.

When Lily begins her painting, at the start of the novel, she is trying to describe the objective facts of her sensation. "What she [Lily] wished to get hold of," Woolf writes, "was that very jar on the nerves, the thing itself before it has been made anything. Get that and start afresh; get that and start afresh; she said desperately, pitching herself firmly against her easel." But that reality — the world seen without a self — is exactly what we can never see. Through the struggle of the artistic process, Lily learns this. "She smiled ironically. For had she not thought, when she began, that she had solved her problem?"

By the end of the novel, Lily knows that her problem has no solution. The self cannot be escaped; reality cannot be unraveled. "Instead," Lily thinks, "there are only little daily miracles, illuminations, matches struck unexpectedly in the dark." Her painting, full as it is of discordant brushstrokes, makes no grand claims. She knows it is only a painting, destined for attics. It will solve nothing, but then, nothing is ever really solved. The real mysteries persist, and "the great revelation never comes." All Lily wants is her painting "to be on a level with ordinary experience, to feel simply that's a chair, that's a table, and yet at the same time, it's a miracle, it's an ecstasy."

And then, with that brave brushstroke down the middle, Lily sees what she wanted to express, even if only for a moment. She does this not by forcing us into some form, but by accepting the fragile reality of our experience. Her art describes us as we are, as a "queer amalgamation of dream and reality, that perpetual marriage of granite and rainbow." Our secret, Lily knows, is that we have no answer. What she does is ask the question. The novel ends on this tonic note of creation:

> With a sudden intensity, as if she saw it clear for a second, she drew a line there, in the centre. It was done; it was finished. Yes, she thought, laying down her brush in extreme fatigue, I have had my vision.

Coda

To say that we should drop the idea of truth as out
there waiting to be discovered is not to say that we
have discovered that, out there, there is no truth.

— Richard Rorty

In 1959, C. P. Snow famously declared that our two cultures — art
and science — suffered from a "mutual incomprehension." As a result, Snow said, our knowledge had become a collection of lonely
fiefdoms, each with its own habits and vocabularies. "Literary intel-
lectuals" analyzed T. S. Eliot and *Hamlet,* while scientists studied
the elementary particles of the universe. "Their attitudes are so different," wrote Snow, "that they can't find much common ground."

Snow's solution to this epistemic schism was the formation of a
"third culture." He hoped that this new culture would close the
"communications gap" between scientists and artists. Each side would
benefit from an understanding of the other, as poets contemplated
Einstein and physicists read Coleridge. Our fictions and our facts
would feed off each other. Furthermore, this third culture would
rein in the extravagances of both cultures at their extremes.

Snow turned out to be prophetic, at least in part. The third culture is now a genuine cultural movement. However, while this new
third culture borrows Snow's phrase, it strays from his project. Instead of referring to a dialogue between artists and scientists — a

shared cultural space, so to speak — the third culture of today refers to scientists who communicate directly with the general public. They are translating their truths for the masses.

On the one hand, this is an important and necessary development. Many of the scientists that make up our current third culture have greatly increased the public's understanding of the scientific avant-garde. From Richard Dawkins to Brian Greene, from Steven Pinker to E. O. Wilson, these scientists do important scientific research and write in elegant prose. Because of their work, black holes, memes, and selfish genes are now part of our cultural lexicon.

Look deeper, however, and it becomes clear that this third culture has serious limitations. For one thing, it has failed to bridge the divide between our two existing cultures. There is still no dialogue of equals. Scientists and artists continue to describe the world in incommensurate languages.

Furthermore, the views promulgated by these scientific thinkers often take a one-dimensional view of the scientific enterprise and its relationship to the humanities. In E. O. Wilson's *Consilience* — a book that is often considered a manifesto for the third culture movement — Wilson argues that the humanities should be "rationalized," their "lack of empiricism" corrected by reductionist science. "The central idea of the consilience world view," Wilson writes, "is that all tangible phenomena, from the birth of stars to the workings of social institutions, are based on material processes that are ultimately reducible, however long and tortuous the sequences, to the laws of physics."

Wilson's ideology is technically true but, in the end, rather meaningless. No serious person denies the reality of gravity or the achievements of reductionism. What Wilson forgets, however, is that not every question is best answered in terms of quantum mechanics. When some things are broken apart, they are just broken. What the artists in this book reveal is that there are many different ways of describing reality, each of which is capable of generating truth. Physics is useful for describing quarks and galaxies, neuroscience is useful for describing the brain, and art is useful for describing our actual experience. While these levels are obviously interconnected, they are also autonomous: art is not reducible to physics. (As Robert Frost wrote, "Poetry is what gets lost in translation.") This is what our third culture *should* be about. It should be a celebration of pluralism.

Unfortunately, many of the luminaries of our current third culture are extremely antagonistic toward everything that isn't scientific. They argue that art is a symptom of our biology, and that anything that isn't experimental is just entertainment. Even worse, our third culture indulges in this attitude without always understanding the art it attempts to encompass. Steven Pinker's book *The Blank Slate: The New Sciences of Human Nature* is a perfect example of this habit.

Pinker sets out to demolish the old intellectual belief in three false constructs: the Blank Slate (the belief that the mind is shaped primarily by its environment), the Noble Savage (the belief that people are naturally good but are ruined by society), and the Ghost in the Machine (the belief that there is a nonbiological entity underlying consciousness). The artists and humanists who promulgate these romantic myths are, of course, the archenemies of the rational evolutionary psychologists and neuroscientists that Pinker defends.

"The giveaway may be found," Pinker writes, "in a now famous statement from Virginia Woolf: 'On or about December 1910, human nature changed.'" For Pinker, Woolf embodies "the philosophy of modernism that would dominate the elite arts and criticism for much of the twentieth century, and whose denial of human nature was carried over with a vengeance to postmodernism." Woolf was wrong, Pinker says, because "human nature did not change in 1910, or in any year thereafter."

Pinker has misunderstood Woolf. She was being ironic. The quote comes from an essay entitled "Character in Fiction," in which she criticizes earlier novelists precisely because they ignored the inner workings of the mind. Woolf wanted to write novels that reflected human nature. She understood, like Pinker, that certain elements of consciousness were constant and universal. Every mind was naturally fragmented, and yet every self emerged from its fragments in the same way. It is this psychological process that Woolf wanted to translate into a new literary form.

But if Pinker is wrong to thoughtlessly attack Virginia Woolf (seeing an enemy when he should see an ally), he is right to admonish what he calls "the priests of postmodernism." Too often, postmodernism — that most inexplicable of -*ism*s — indulges in cheap disavowals of science and the scientific method. There is no truth, postmodernists say, only differing descriptions, all of which are equally invalid. Obviously, this idea very quickly exhausts itself. No truth is perfect, but that doesn't mean all truths are equally *im*perfect. We will always need some way to distinguish among our claims.

Thus, in our current culture, we have two epistemological extremes reflexively attacking the other. Postmodernists have ignorantly written off science as nothing but another text, and many scientists have written off the humanities as hopelessly false. Instead of constructing a useful dialogue, our third culture has only inflamed this sad phenomenon.

Before she began writing *Mrs. Dalloway,* Virginia Woolf wrote that in her new novel the "psychology should be done very realistically." She wanted this book to capture the mind in its actual state, to express the tumultuous process at the center of our existence. For too long, Woolf believed, fiction had indulged in a simplified view of consciousness. She was determined to make things complicated.

The artistic exploration of the mind did not end with Virginia Woolf. In 2005, the British novelist Ian McEwan gave *Mrs. Dalloway* — a novel set on a single day in the life of an upper-class Londoner — a scientific update. His novel *Saturday* reworks the narrative structure of Woolf (which was itself a reworking of *Ulysses*), but this time from the perspective of a neurosurgeon. As a result, the psychology is done *very* realistically. Like *Mrs. Dalloway, Satur-*

day is shadowed by war and madness, and includes oblique references to planes in flight and the prime minister. The ordinary moments of life — from grocery shopping to games of squash — are shown to contain *all* of life.

Saturday begins before dawn. The protagonist, Dr. Henry Perowne, finds himself awake, although "it's not clear to him when exactly he became conscious, nor does it seem relevant." All he knows is that his eyes are open, and that he *exists,* a palpable, if immaterial, presence: "It's as if, standing there in the darkness, he's materialized out of nothing."

Of course, being a neurosurgeon, Henry knows better. He is on intimate terms with the cortex. It is "a kind of homeland to him." He believes that the mind is the brain, and that the brain is just a myelinated mass of fissures and folds. McEwan, who spent more than two years following around a neurosurgeon during the writing of the book, delights in the odd consonances of our anatomy. He insists on showing us the sheer strangeness of our source.

But McEwan simultaneously complicates the materialist world his character inhabits. Though Henry disdains philosophy and is bored by fiction, he is constantly lost in metaphysical reveries. As he picks up fish for dinner, Henry wonders "what the chances are, of this particular fish, from that shoal, ending up in the pages, no, on this page of this copy of the *Daily Mirror?* Something just short of infinity to one. Similarly, the grains of sand on a beach, arranged just so. The random ordering of the world, the unimaginable odds against any particular condition." And yet, despite the odds, our reality holds itself together: the fish is there, wrapped in newspaper in the plastic bag. Existence is a miracle.

It is also a precarious miracle. Woolf showed us this with Septimus, whose madness served to highlight the fragility of sanity. McEwan chooses Baxter, a man suffering from Huntington's disease, to produce a parallel effect. Baxter's disease, thinks the neurosurgeon, "is biological determinism in its purest form. The misfortune lies within a single gene, in an excessive repeat of a single sequence — CAG." There is no escape from this minor misprint.

But McEwan doesn't make the logical mistake of believing that such a deterministic relationship is true of life in general. Henry knows that the real gift of our matter is to let us be *more* than matter. While operating on an exposed brain, Henry ruminates on the mystery of consciousness. He knows that even if science "solves" the brain, "the wonder will remain. That mere wet stuff can make this bright inward cinema of thought, of sight and sound and touch bound into a vivid illusion of an instantaneous present, with a self, another brightly wrought illusion, hovering like a ghost at its center. Could it ever be explained, how matter becomes conscious?"

Saturday does not answer the question. Instead, the novel strives to remind us, again and again, that the question has no answer. We will never know how the mind turns the water of our cells into the wine of consciousness. Even Baxter, a man defined by his tragic genetic flaw, is ultimately altered by a poem. When Henry's daughter begins reciting Matthew Arnold's "Dover Beach," a poem about the melancholy of materialism, Baxter is transfixed. The words "touched off a yearning he could barely begin to define." The plot of *Saturday* hinges on this chance event, on a mind being moved by nothing more real than rhyming words. Poetry sways matter. Could anything be less likely?

McEwan ends *Saturday* the way he began it: in the dark, in the present tense, with Henry in bed. It has been a long day. As Henry is drifting off to sleep, his last thoughts are not about the brain, or surgery, or materialism. All of that seems far away. Instead, Henry's thoughts return to the only reality we will ever know: our *experience*. The feeling of consciousness. The feeling of feeling. "There's always this, is one of his remaining thoughts. And then: there's only this."

McEwan's work is a potent demonstration that even in this age of dizzying scientific detail, the artist remains a necessary voice. Through the medium of fiction, McEwan explores the limits of science while doing justice to its utility and eloquence. Though he never doubts our existence as a property of matter — this is why the surgeon can heal our wounds — McEwan captures the paradox of being a mind aware of itself. While each self is a brain, it is a brain that contemplates its own beginnings.

Saturday is a rare cultural commodity, and not only because of McEwan's artistry. It symbolizes, perhaps, the birth of a new fourth culture, one that seeks to discover relationships *between* the humanities and the sciences. This fourth culture, much closer in concept to Snow's original definition (and embodied by works like *Saturday*), will ignore arbitrary intellectual boundaries, seeking instead to blur the lines that separate. It will freely transplant knowledge between the sciences and the humanities, and will focus on connecting the reductionist fact to our actual experience. It will take a pragmatic view of the truth, and it will judge truth not by its origins but in terms of its usefulness. What does this novel or experiment or poem or protein teach us about ourselves? How does it help us to understand who we are? What long-standing problem has it solved?

If we are open-minded in our answers to these questions, we will discover that the poem can be just as true and useful as the acronym. And while science will always be our primary method of investigating the universe, it is naïve to think that science can solve everything by itself, or that everything can even be solved. One of the ironies of modern science is that some of its most profound discoveries — like Heisenberg's uncertainty principle,* or the emergent nature of consciousness — are actually about the limits of science. As Vladimir Nabokov, the novelist and lepidopterist, once put it, "The greater one's science, the deeper the sense of mystery."

We now know enough to know that we will never know everything. This is why we need art: it teaches us to how live with mystery. Only the artist can explore the ineffable without offering us an answer, for sometimes there is no answer. John Keats called this romantic impulse "negative capability." He said that certain poets, like

---

* This principle of quantum physics states that one can know either the position of a particle or its momentum (mass times velocity), but not both variables simultaneously. In other words, we can't know everything about anything.

Shakespeare, had "the ability to remain in uncertainties, mysteries, doubts, without any irritable reaching after fact and reason." Keats realized that just because something can't be solved, or reduced into the laws of physics, doesn't mean it isn't real. When we venture beyond the edge of our knowledge, all we have is art.

But before we can get a fourth culture, our two existing cultures must modify their habits. First of all, the humanities must sincerely engage with the sciences. Henry James defined the writer as someone on whom nothing is lost; artists must heed his call and not ignore science's inspiring descriptions of reality. Every humanist should read *Nature*.

At the same time, the sciences must recognize that their truths are not the only truths. No knowledge has a monopoly on knowledge. That simple idea will be the starting premise of any fourth culture. As Karl Popper, an eminent defender of science, wrote, "It is imperative that we give up the idea of ultimate sources of knowledge, and admit that all knowledge is human; that it is mixed with our errors, our prejudices, our dreams, and our hopes; that all we can do is to grope for truth even though it is beyond our reach. There is no authority beyond the reach of criticism."

I hope that this book has shown how art and science might be reintegrated into an expansive critical sphere. Both art and science can be useful, and both can be true. In our own time, art is a necessary counterbalance to the glories and excesses of scientific reductionism, especially as they are applied to human experience. This is the artist's purpose: to keep *our* reality, with all its frailties and question marks, on the agenda. The world is large, as Whitman once remarked. It contains multitudes.

Acknowledgments

WHERE TO BEGIN? I guess at this book's beginning. My agent, Emma Parry, got in contact with me after reading a short article of mine in *Seed* magazine. Her patience and guidance were extraordinary. It took many incoherent drafts, but she eventually helped me shape my disorganized thoughts about Proust and neuroscience into a coherent book proposal.

I did most of my research at Oxford University. For giving me the luxury of spending all day in a library, I thank the Rhodes Trust.

I first had the idea for this book while working in the neuroscience lab of Dr. Eric Kandel. When I began working in Dr. Kandel's lab, I dreamed of being a scientist. After spending a few years with his postdocs and grad students, I quickly realized that I wasn't good enough. (W. H. Auden once said that when he found himself in the company of scientists he felt like "a shabby curate who has strayed by mistake into a room full of dukes." I understand the feeling entirely.) I will always be grateful to Dr. Kandel for giving me the chance to participate in the scientific process. In addition, I want to thank all the scientists who talked with me while I wrote this book. I'm convinced that postdocs are the suffering artists of the twenty-first century.

In Dr. Kandel's lab, I worked for Dr. Kausik Si. I could not have asked for a better mentor. His brilliance and kindness will always inspire me. Although I probably set his research back a few years — I excelled at experimental failure — our conversations about literature and science were essential to my thinking.

And then there are those patient souls who read my rough drafts and told me how to make them better. This book would be incomparably worse were it not for Jad Abumrad, Steven Pulimood, Paul Tullis, Sari Lehrer, and Robert Krulwich. I would also like to thank everyone at Houghton Mifflin, especially Will Vincent, who ably led me through the logistics of creating a book, and Tracy Roe, who did a fantastic job of finding and fixing errors in the manuscript.

Amanda Cook is a truly amazing editor. I really can't say enough good things about her. If it were not for her comments, this book would be twice as long and half as readable. She not only fixed my prose, she fixed my ideas. Amanda also trusted me to write about a very big subject, and for that I will always be grateful.

Finally, I owe an impossible debt to my girlfriend, Sarah Liebowitz, and my mother, Ariella Lehrer. They both have read this book way too many times to count. When I was upset with my writing, they cheered me up. And when I forgot how much work remained, they reminded me. Every sentence is better because of their support, criticism, and love. Without whom not.

Notes

1. Walt Whitman: The Substance of Feeling

1 *"Was somebody asking"*: Walt Whitman, *Leaves of Grass: The "Death-Bed" Edition* (New York: Random House, 1993), 27.

2 *"a person"*: Ibid., 702.

"Your very flesh": As cited in Paul Berman, "Walt Whitman's Ghost," *The New Yorker* (June 12, 1995): 98–104.

"The mind is embodied": Antonio Damasio, *Descartes' Error* (London: Quill, 1995), 118.

4 *"You might as easily"*: Brian Burrell, *Postcards from the Brain Museum* (New York: Broadway Books, 2004), 211.

"the greatest conglomeration": Jerome Loving, *Walt Whitman* (Berkeley: University of California Press, 1999), 104.

5 *"its totality"*: Horace Traubel, *Intimate with Walt: Selections from Whitman's Conversations with Horace Traubel, 1882–1892* (Des Moines: University of Iowa Press, 2001).

"I was simmering": Loving, *Walt Whitman*, 168.

7 *"I like the silent"*: Ralph Waldo Emerson, *Nature, Addresses, and Lectures* (Boston: Houghton Mifflin, 1890), 272.

"broad shouldered": Loving, *Walt Whitman*, 224.

"Leading traits of character": Ibid., 150.

"one of the richest": Donald D. Kummings and J. R. LeMaster, eds., *Walt Whitman: An Encyclopedia* (New York: Garland, 1998), 206.

"The poet stands among": Emerson, *Nature*, 455.

8 *"Doubt not"*: Ralph Waldo Emerson, *Selected Essays, Lectures, and Poems* (New York: Bantam, 1990), 223.

9 *"I am the poet"*: Ed Folsom and Kenneth M. Price, "Biography," Walt Whit-

man Archive, http://www.whitmanarchive.org/biography (accessed January 7, 2005).

"I am not blind": Loving, *Walt Whitman*, 189.

10 *"in danger"*: Ibid., 241.

"What does a man": Ibid.

"the human body": Whitman, *Leaves of Grass: The "Death-Bed" Edition*, 699.

"O my body": Ibid., 128.

11 *"where their priceless blood"*: Ibid., 387.

"pieces of barrel-staves": Loving, *Walt Whitman*, 1.

"the heap of feet": Edwin Haviland Miller, ed., *Walt Whitman: The Correspondence.* (New York: New York University Press, 1961–1977), 59.

"Those three": Ibid., 77.

12 *"opened a new world"*: Loving, *Walt Whitman*, 1.

"From the stump": Whitman, *Leaves of Grass: The "Death-Bed" Edition*, 388.

"the hiss of the surgeon's knife": Ibid., 91.

13 *"a sense of the existence"*: Silas Weir Mitchell, *Injuries of Nerves, and Their Consequences* (Philadelphia: Lippincott, 1872).

"invisibly and uninterpenetratingly": Herman Melville, *Redburn, White-Jacket, Moby Dick* (New York: Library of America, 1983), 1294–98.

14 *"to his horror"*: Laura Otis, ed., *Literature and Science in the Nineteenth Century* (Oxford: Oxford University Press, 2002), 358–63.

15 *"How much of the limb"*: William James, "The Consciousness of Lost Limbs," *Proceedings of the American Society for Psychical Research* 1 (1887).

16 *"passionate and mystical"*: William James, *Writings: 1878–1899* (New York: Library of America, 1987), 851.

"contemporary prophet" : Bruce Wilshire, ed., *William James: The Essential Writings* (Albany: State University of New York, 1984), 333.

"the kind of fiber": Ibid., 337.

17 *"It is a sort of work"*: Louis Menand, *The Metaphysical Club* (New York: Farrar, Straus and Giroux, 2001), 324.

18 *"The demand for atoms"*: James, *Writings: 1878–1899*, 996.

"I will not make poems": Whitman, *Leaves of Grass: The "Death-Bed" Edition*, 26.

"there would be nothing": William James, "What Is an Emotion?" *Mind* 9 (1884): 188–205.

20 *"The body contributes"*: Antonio Damasio, *Descartes' Error* (London: Quill, 1995), 226.

One of Damasio's most surprising discoveries: Ibid., 212–17.

"There is more reason": Friedrich Nietzsche, *The Portable Nietzsche* (Viking: New York, 1977), 146.

22 *"the curious sympathy"*: Whitman, *Leaves of Grass: The "Death-Bed" Edition*, 130.

"the spirit receives": Walt Whitman, *Leaves of Grass* (Oxford: Oxford University Press, 1998), 456.

"I will make the poems": Ibid., 20.

"Hurray for positive science": Whitman, *Leaves of Grass: The "Death-Bed" Edition,* 64.

"for they are vascular": Ralph Waldo Emerson, *Selected Essays, Lectures, and Poems* (New York: Bantam, 1990), 291.

23 *"I and this mystery":* Whitman, *Leaves of Grass: The "Death-Bed" Edition,* 36.

"When you organize": Randall Jarrell, *No Other Book* (New York: HarperCollins, 1999), 118.

24 *"Now I see it is true":* Whitman, *Leaves of Grass: The "Death-Bed" Edition,* 77.

2. George Eliot: The Biology of Freedom

25 *"Seldom, very seldom":* Jane Austen, *Emma* (New York: Modern Library, 1999), 314.

"simply a set of experiments": Gordon Haight, ed., *George Eliot's Letters* (New Haven: Yale University Press, 1954–1978), vol. VI, 216–17.

26 *Henry James once:* David Caroll, ed., *George Eliot: The Critical Heritage* (London: Routledge and Kegan Paul, 1971), 427.

27 *"had no need":* Michael Kaplan and Ellen Kaplan, *Chances Are . . .* (New York: Viking, 2006), 42.

29 *"We must . . . imagine":* Louis Menand, *The Metaphysical Club* (New York: Farrar, Straus and Giroux, 2002), 195.

31 *"I want to know":* Haight, ed., *George Eliot's Letters,* vol. VIII, 56–5.

"The lack of physical": Ibid., 43.

32 *"She might have compared":* George Eliot, *Middlemarch* (London: Norton, 2000), 305.

"We are not 'judicious'": Valerie A. Dodd, *George Eliot: An Intellectual Life* (London: Macmillan, 1990), 227.

"a scientific poet": George Levine, ed., *Cambridge Companion to George Eliot* (Cambridge: Cambridge University Press, 2001), 107.

33 *"no thinking man":* George Lewes, *Comte's Philosophy of Science* (London: 1853), 92.

"Necessitarianism": Haight, ed., *George Eliot's Letters,* vol. IV, 166.

34 *"Was she beautiful":* Eliot, *Daniel Deronda* (New York: Penguin Classics, 1996), 1.

"epoch": Haight, ed., *George Eliot's Letters,* vol. III, 214.

35 *"Even Science":* Eliot, *Daniel Deronda,* 1.

36 *"If we had a keen":* Eliot, *Middlemarch,* 124.

"There lives more faith": Rosemary Ashton, *George Eliot: A Life* (New York: Allen Lane, 1996), 145.

"lost among small closets": Eliot, *Middlemarch,* 126.

37 *"hard, unaccommodating Actual"*: Eliot, *Daniel Deronda*, 380.
"There is no creature": Eliot, *Middlemarch*, 514.
"Every limit": Ibid., 512.
"to throw the whole": As cited in Ashton, *George Eliot: A Life*, 305.

38 *"is not cut"*: Eliot, *Middlemarch*, 734.
"We are conscious": Thomas Huxley, "On the Hypothesis That Animals Are Automata, and Its History," *Fortnightly Review* (1874): 575–77.

40 *Beginning in 1962:* J. Altman, "Are New Neurons Formed in the Brains of Adult Mammals?" *Science* 135 (1962): 1127–28.
Kaplan discovered: M. S. Kaplan, "Neurogenesis in the Three-Month-Old Rat Visual Cortex," *Journal of Comparative Neurology* 195 (1981): 323–38.
"At the time": Personal interview at Rockefeller Field Research Center, July 26, 2006.

41 *"Take nature away"*: Michael Specter, "Rethinking the Brain," *The New Yorker*, July 23, 2001.
"Until the scientist": Thomas Kuhn, *The Structure of Scientific Revolutions*, 3rd ed. (Chicago: University of Chicago Press, 1996), 53.

42 *monkey mothers who live in stressful:* C. L. Coe, et al., "Prenatal Stress Diminishes Neurogenesis in the Dentate Gyrus of Juvenile Rhesus Monkeys," *Biology of Psychiatry* 10 (2003): 1025–34.
The hippocampus: F. H. Gage et al., "Survival and Differentiation of Adult Neural Progenitor Cells Transplanted to the Adult Brain," *Proceedings of the National Academy of Sciences* 92 (1995).
which help us to learn: Greg Miller, "New Neurons Strive to Fit In," *Science* 311 (2006): 938–40.

43 *antidepressants work:* Luca Santarelli et al., "Requirement of Hippocampal Neurogenesis for the Behavioral Effects of Antidepressants," *Science* 301 (2003): 805–08.

44 *"Once 'information' has passed"*: Robert Olby, *The Path to the Double Helix* (London: Macmillan, 1974), 432.
"We are survival machines": Richard Dawkins, *The Selfish Gene* (Oxford: Oxford University Press, 1976), ix.

45 *the code sequence:* Richard Lewontin, *Biology as Ideology* (New York: Harper Perennial, 1993), 67.

46 *their auditory cortex now:* J. Sharma, A. Angelucci, and M. Sur, "Induction of Visual Orientation Modules in Auditory Cortex," *Nature* 404 (2000): 841–47.
"the most compelling": Sandra Blakeslee, "Rewired Ferrets Overturn Theories of Brain Growth," *New York Times*, April 25, 2000, sec. F1.

50 *Gage's new hypothesis:* A. R. Muotri et al., "Somatic Mosaicism in Neuronal Precursor Cells Mediated by L1 Retrotransposition," *Nature* 435: 903–10.
"The more diversified": Charles Darwin, *On the Origin of Species by Means of*

Natural Selection, or the Preservation of Favored Races in the Struggle for Life (London: John Murray, 1859), 112.

51 *"highly irregular"*: Karl Popper, *Objective Knowledge* (Oxford: Oxford University Press, 1972), ch. 6.

"Art is the nearest": George Eliot, "The Natural History of German Life," *Westminster Review,* July 1856.

"I refuse": Haight, ed., *George Eliot's Letters,* vol. VI, 216–17.

52 *"I shall not"*: Ibid., 166.

3. Auguste Escoffier: The Essence of Taste

54 *"Indeed, stock is everything"*: Auguste Escoffier, *The Escoffier Cookbook: A Guide to the Fine Art of Cookery for Connoisseurs, Chefs, Epicures* (New York: Clarkson Potter, 1941), 1.

"pleasurable occasion": Auguste Escoffier, *Escoffier: The Complete Guide to the Art of Modern Cookery* (New York: Wiley, 1983), xi.

55 *"carefully studying"*: Amy Trubek, *Haute Cuisine: How the French Invented the Culinary Profession* (Philadelphia: University of Pennsylvania Press, 2001), 126.

"Experience alone": Escoffier, *The Escoffier Cookbook,* 224.

56 *"The discovery of"*: Jean Anthelme Brillat-Savarin, trans. M.F.K. Fisher, *The Physiology of Taste* (New York: Counterpoint Press, 2000), 4.

57 *"There is a taste"*: Alex Renton, "Fancy a Chinese?" *Observer Food Magazine,* July 2005, 27–32.

58 *"This study has"*: Ibid.

The chemical acronym: K. Ikeda, "New Seasonings," *Journal of the Chemical Society of Tokyo* 30 (1909): 820–36.

according to Democritus: J. I. Beare, ed., *Greek Theories of Elementary Cognition from Alcmaeon to Aristotle* (Oxford: Clarendon Press, 1906), 164.

In De Anima: Stanley Finger, *Origins of Neuroscience* (Oxford: Oxford University Press, 1994), 165.

60 *As a leg of prosciutto ages:* For a delightful tour of the culinary uses of umami see Jeffrey Steingarten, *It Must've Been Something I Ate* (New York: Vintage, 2003), 85–99.

61 *The first receptor was discovered:* N. Chaudhari et al., "A Novel Metabotropic Receptor Functions as a Taste Receptor," *Nature Neuroscience* 3 (2000): 113–19.

The second sighting: G. Nelson et al., "An Amino-Acid Taste Receptor," *Nature* 416 (2002): 199–202.

62 *Unlike the tastes:* M. Schoenfeld et al., "Functional MRI Tomography Correlates of Taste Perception in the Human Primary Taste Cortex," *Neuroscience* 127 (2004): 347–53.

breast milk has ten times: Stephen Pincock, "All in Good Taste," *FT Magazine,* June 25, 2005, 13.

"A well-displayed meal": Kenneth James, *Escoffier: The King of Chefs* (London: Hambledon and London, 2002), 109.

63 *"The customer":* Escoffier, *The Complete Guide,* 67.

66 *"No matter how high":* Richard Axel, lecture, December 1, 2005: MIT, Picower Institute.

68 *Our sense of smell:* Rachel Herz, "The Effect of Verbal Context on Olfactory Perception," *Journal of Experimental Psychology: General* 132 (2003): 595–606.

This feedback continually: Eric Kandel, James Schwartz, and Thomas Jessell, *Principles of Neural Science,* 4th ed. (New York: McGraw Hill, 2000), 632.

But when that same air: I. E. de Araujo et al., "Cognitive Modulation of Olfactory Processing," *Neuron* 46 (2005): 671–79.

69 *"Even horsemeat":* James, *Escoffier: The King of Chefs,* 47.

70 *"can be much more":* Daniel Zwerdling, "Shattered Myths," *Gourmet* (August 2004), 72–74.

71 *"organizing system":* Donald Davidson, *Inquiries into Truth and Interpretation* (Oxford: Oxford University Press, 2001), 189.

Everything else withers: O. Beluzzi et al., "Becoming a New Neuron in the Adult Olfactory Bulb," *Nature Neuroscience* 6 (2003): 507–18.

72 *Subjects repeatedly exposed:* J. D. Mainland et al., "One Nostril Knows What the Other Learns," *Nature* 419 (2002): 802.

What was once: C. J. Wysocki, "Ability to Perceive Androstenone Can Be Acquired by Ostensibly Anosmic People," *Proceedings of the National Academy of Sciences* 86 (1989). And L. Wang et al., "Evidence for Peripheral Plasticity in Human Odour Response," *Journal of Physiology* (January 2004): 236–44.

73 *He refused to learn:* James, *Escoffier: The King of Chefs,* 132.

"No theory": Escoffier, *The Escoffier Cookbook,* 1.

74 *"If it works":* Sam Sifton, "The Cheat," *New York Times Magazine,* May 8, 2005.

4. Marcel Proust: The Method of Memory

75 *"Even a bureau":* Charles Baudelaire, *Baudelaire in English* (New York: Penguin, 1998), 91.

76 *"for if our life":* Marcel Proust, *Time Regained,* vol. VI (New York: Modern Library, 1999), 441.

78 *"recognize in his own":* Ibid., 322.

"I have enough": As cited in Joshua Landy, *Philosophy as Fiction: Self, Deception, and Knowledge in Proust* (Oxford: Oxford University Press, 2004), 163.

79 *"The kind of literature":* Proust, *Time Regained,* 284.

"the structure of his spirit": Ibid., 206.

"No sooner had": Marcel Proust, *Swann's Way*, vol. I (New York: Modern Library, 1998), 60.

80 *"When from a long"*: Ibid., 63.

Rachel Herz: Rachel Herz and J. Schooler, "A Naturalistic Study of Autobiographical Memories Evoked by Olfactory and Visual Cues: Testing the Proustian Hypothesis," *American Journal of Psychology* 115 (2002): 21–32.

81 *"Perhaps because I had"*: Proust, *Swann's Way*, 63.

"the game wherein": Ibid., 64.

"It is a labor in vain": Ibid., 59.

82 *"I am obliged to depict errors"*: As cited in Landy, *Philosophy as Fiction*, 4.

83 *"speculative cavort"*: Stanley Finger, *Minds Behind the Brain* (Oxford: Oxford University Press, 2000), 214.

84 *But in a set of extraordinary*: Karim Nader et al., "Fear Memories Require Protein Synthesis in the Amygdala for Reconsolidation after Retrieval," *Nature* 406: 686–87. See also J. Debiec, J. LeDoux, and K. Nader, "Cellular and Systems Reconsolidation in the Hippocampus," *Neuron* 36 (2002); and K. Nader et al., "Characterization of Fear Memory Reconsolidation," *Journal of Neuroscience* 24 (2004): 9269–75.

89 *"In this book"*: Proust, *Time Regained*, 225.

"How paradoxical it is": Proust, *Swann's Way*, 606.

91 *This theory, published in 2003*: K. Si, E. Kandel, and S. Lindquist, "A Neuronal Isoform of the Aplysia CPEB Has Prion-Like Properties," *Cell* 115 (2003): 879–91.

It is at these tiny crossings: Kelsey Martin et al., "Synapse-Specific, Long-Term Facilitation of Aplysia Sensory to Motor Synapses: A Function for Local Protein Synthesis in Memory Storage," *Cell* 91 (1997): 927–38.

92 *He had heard of a molecule*: Joel Richter, "Think Globally, Translate Locally: What Mitotic Spindles and Neuronal Synapses Have in Common," *Proceedings of the National Academy of Sciences* 98 (2001): 7069–71.

This same molecule also happened: L. Wu et al., "CPEB-Mediated Cytoplasmic Polyadenylation and the Regulation of Experience-Dependent Translation of Alpha-CaMKII mRNA at Synapses," *Neuron* 21 (1998): 1129–39.

94 *Essentially, the more likely*: A. Papassotiropoulos et al., "The Prion Gene Is Associated with Human Long-Term Memory," *Human Molecular Genetics* 14 (2005): 2241–46.

Other experiments have linked: J. M. Alarcon et al., "Selective Modulation of Some Forms of Schaffer Collateral-CA1 Synaptic Plasticity in Mice with a Disruption of the CPEB-1 Gene," *Learning and Memory* 11 (2004): 318–27.

95 *"The past is hidden"*: Proust, *Swann's Way*, 59.

5. Paul Cézanne: The Process of Sight

96 *"that on or about"*: Virginia Woolf, *Collected Essays* (London: Hogarth Press, 1966–1967), vol. I, 320.

"art aim at a pseudo-scientific": Vassiliki Kolocotroni, Jane Goldman, and Olga Taxidou, eds., *Modernism: An Anthology of Sources and Documents* (Chicago: University of Chicago Press, 1998), 189–92.

97 *"being of no interest"*: Christopher Butler, *Early Modernism* (Oxford: Oxford University Press, 1994), 216.

"The eye is not": Ulrike Becks-Malorny, *Cézanne* (London: Taschen, 2001), 46.

100 *"to be the servant"*: Charles Baudelaire, *Charles Baudelaire: The Mirror of Art*, trans. Jonathan Mayne (London: Phaidon Press, 1955).

"the transient": Charles Baudelaire, *Baudelaire: Selected Writings on Art and Artists*, trans. P. E. Charvet (Cambridge: Cambridge University Press, 1972).

102 *"I want to make"*: John Rewald, *Cézanne* (New York: Harry Abrams, 1986), 159.

"Teachers are all": Peter Schjeldahl, "Two Views," *The New Yorker*, July 11, 2005.

"The eye must absorb everything": Becks-Malorny, *Cézanne*, 24.

104 *"I tried to copy nature"*: Michael Doran, ed., *Conversations with Cézanne* (Berkeley: University of California Press, 2001), 120.

107 *"it is as if there"*: As cited in Daniel Schwarz, *Reconfiguring Modernism* (New York: Palgrave Macmillan, 1997), 108.

108 *Why does the mind*: M. Bar et al., "Top-Down Facilitation of Visual Recognition," *Proceedings of the National Academy of Sciences* 103 (2006): 449–54.

"His [Dr. P's] responses": Oliver Sacks, *The Man Who Mistook His Wife for a Hat* (London: Picador, 1985), 9.

109 *"He [Dr. P] then started"*: Ibid., 10.

110 *"Is painting only a whim"*: Becks-Malorny, *Cézanne*, 8.

112 *"employing the experimental"*: Kolocotroni, *Modernism*, 170.

"wild mental activity": Emile Zola, trans. Thomas Walton, *The Masterpiece* (Oxford: Oxford University Press, 1999), x.

"describe man": Ibid., 180.

"new literature": Ibid.

"Our enemies": Rachel Cohen, "Artist's Model," *The New Yorker*, November 7, 2005, 62–85.

113 *"disappear, and simply show"*: Kolocotroni, *Modernism*, 173.

"I have a better understanding": Rewald, *Cézanne*, 182.

116 *"The imagination"*: Immanuel Kant, *The Critique of Pure Reason*, trans. J.M.D. Meiklejohn (New York: Prometheus Books, 1990).

117 *Unlike the Wundtians*: Mitchell G. Ash, *Gestalt Psychology in German Cul-*

ture, 1890–1967: Holism and the Quest for Objectivity (Cambridge: Cambridge University Press, 1998), 126.

The visual cortex is divided: The V5 region is often referred to as the MT region. Richard Born and D. Bradley, "Structure and Function of Visual Area MT," *Annual Review of Neuroscience* 28 (2005): 157–89.

118 *When these specific neurons light up:* R. Quiroga et al., "Invariant Visual Representation by Single Neurons in the Human Brain," *Nature* 435 (2005): 1102–07.

"*A sensation is rather like*": William James, *Writings 1902–1910* (New York: Library of America, 1987), 594.

119 *His paintings are criticisms:* The critic Clement Greenberg argued that such Kantian self-criticism represented the essence of modernism. The modernist artist, he said, uses "the characteristic methods of a discipline to criticize the discipline itself." Clement Greenberg, "Modernist Painting," *Art and Literature* 4 (1965).

"*Cézanne made the fruit*": Rainer Maria Rilke, *Letters on Cézanne* (London: Vintage, 1985), 33.

6. Igor Stravinsky: The Source of Music

121 "*the immense sensation*": Stephen Walsh, *Igor Stravinsky: A Creative Spring* (Berkeley: University of California, 2002), 208.

122 "*We could hear nothing*": Gertrude Stein, *The Autobiography of Alice B. Toklas* (London: Penguin Classics, 2001), 150.

123 "*Exactly what I wanted*": Igor Stravinsky and Robert Craft, *Conversations with Igor Stravinsky* (London: Faber, 1979), 46–47.

"*before the arising of Beauty*": Vera Stravinsky and Robert Craft, eds., *Stravinsky in Pictures and Documents* (New York: Simon and Schuster, 1978), 524–26.

125 "*as a period of waiting*": Igor Stravinsky and Robert Craft, *Memories and Commentaries* (London: Faber and Faber, 1960), 26.

127 "*If I must commit*": Alex Ross, "Whistling in the Dark," *The New Yorker,* February 18, 2002.

"*The overwhelming multitude*": As cited in Charles Rosen, *Arnold Schoenberg* (Chicago: University of Chicago Press, 1996), 33.

"*Schoenberg is one*": Walsh, *Igor Stravinsky,* 190.

"*One atom of hydrogen*": As cited in Peter Conrad, *Modern Times, Modern Places* (New York: Knopf, 1999), 85.

128 "*If it is art*": Ross, "Whistling in the Dark."

"*rationalism and rules*": Walsh, *Igor Stravinsky,* 397.

129 *When these selective neurons:* D. Bendor and Q. Wang, "The Neuronal Representation of Pitch in the Primate Auditory Cortex," *Nature* 436 (2005): 1161–65.

130 *It projects imaginary order:* A. Patel and E. Balaban, "Temporal Patterns of Human Cortical Activity Reflect Tone Sequence Structure," *Nature* 404 (2002).

131 *Meyer wanted to show:* Leonard Meyer, *Emotion and Meaning in Music* (Chicago: University of Chicago Press, 1961), 145–60.

132 *"For the human":* Ibid., 16.
   "is the whole": Ibid., 151.

133 *"all art aspires":* William Pater, *The Renaissance: Studies in Art and Poetry* (Oxford: Oxford University Press, 1998), 86.

134 *"Very little tradition":* Igor Stravinsky and Robert Craft, *Expositions and Development* (London: Faber and Faber, 1962), 148.

135 *"I have confided":* Stravinsky, ed., *Stravinsky in Pictures,* 524–26.

136 *The order is our own:* Peter Hill, *Stravinsky: The Rite of Spring* (Cambridge: Cambridge University Press, 2000), 53.

137 *"To listen is an effort":* Alex Ross, "Prince Igor," *The New Yorker,* November 6, 2000.
   "Music is": Igor Stravinsky, *Chronicle of My Life* (Gollancz: London, 1936), 91.

138 *"only by being":* Igor Stravinsky, *Poetics of Music* (New York: Vintage, 1947), 23.
   "an ally in the fight": Plato, *Timaeus,* trans. Donald Zeyl (New York: Hackett, 2000), 36.

139 *"In music, advance":* Stravinsky and Craft, *Expositions and Development,* 138.
   "Our little Igor": Walsh, *Igor Stravinsky,* 233.

140 *The brain tunes:* N. Sugaand E. Gao, "Experience-Dependent Plasticity in the Auditory Cortex and the Inferior Colliculus of Bats: Role of the Cortico-Fugal System," *Proceedings of the National Academy of Sciences* 5 (2000): 8081–86.
   Feedback from higher-up: S. A. Chowdury and N. Suga, "Reorganization of the Frequency Map of the Auditory Cortex Evoked by Cortical Electrical Stimulation in the Big Brown Bat," *Journal of Neurophysiology* 83, no. 4 (2000): 1856–63.
   This learning is largely: S. Bao, V. Chan, and M. Merzenich, "Cortical Remodeling Induced by Activity of Ventral Tegmental Dopamine Neurons," *Nature* 412 (2001): 79–84.

141 *In fact, the brainstem:* D. Perez-Gonzalez et al., "Novelty Detector Neurons in the Mammalian Auditory Midbrain," *European Journal of Neuroscience* 11 (2005): 2879–85.
   When the musical pattern: J. R. Hollerman and W. Schultz, "Dopamine Neurons Report an Error in the Temporal Prediction of Reward During Learning," *Nature Neuroscience* 1 (1998): 304–09.
   If dopamine neurons can't correlate: S. Tanaka, "Dopaminergic Control of

Working Memory and Its Relevance to Schizophrenia: A Circuit Dynamics Perspective," *Neuroscience* 139 (2005): 153–71.

Pierre Monteux, the conductor: Thomas Kelly, *First Nights* (New Haven: Yale University Press, 2000), 277.

"Gentlemen, you do not": Kelly, *First Nights,* 281.

142 *"If something is to stay":* Friedrich Nietzsche, *The Genealogy of Morals,* trans. Walter Kaufmann (New York: Vintage, 1989), 61.

7. Gertrude Stein: The Structure of Language

145 *"There are automatic movements":* As cited in Steven Meyer, *Irresistible Dictation* (Stanford: Stanford University Press, 2001), 221.

"There is no good nonsense": Ibid., 228.

Even Stein later admitted: Gertrude Stein, *The Autobiography of Alice B. Toklas* (London: Penguin Classics, 2001), 86.

146 *"A CARAFE, THAT IS":* Gertrude Stein, *The Selected Writings of Gertrude Stein* (New York: Vintage, 1990), 461.

147 *"If an unusual foreign":* William James, *The Principles of Psychology* (New York: Dover, 1950), vol. 1, 262.

"the way sentences diagram": Gertrude Stein, *Lectures in America* (Boston: Beacon Press, 1985), 211.

"Other things may be": Ibid.

148 *"Language is an instinctive":* Charles Darwin, *The Descent of Man and Selection in Relation to Sex* (New York: Hurst and Co., 1874), 101.

"Practical medicine did not": As cited in Meyer, *Irresistible Dictation,* 55.

"and she was bored": Stein, *Autobiography,* 81.

"Either I am crazy": James Mellow, *Charmed Circle: Gertrude Stein and Company* (London: Phaidon, 1974), 45.

149 *"I was alone":* Gertrude Stein, *Picasso* (Boston: Beacon Press, 1959).

They discussed art and philosophy: Judith Ryan, *The Vanishing Subject* (Chicago: University of Chicago Press, 1991), 92.

"Picasso had never": Stein, *Autobiography,* 52–60.

152 *"To see the things":* Mellow, *Charmed Circle,* 430.

153 *"You see":* Robert Haas, ed., *A Primer for the Gradual Understanding of Gertrude Stein* (Los Angeles: Black Sparrow, 1971), 15.

"Henry James was": Stein, *Autobiography,* 87.

"has not one window": As quoted in Jacques Barzun, *A Stroll with William James* (Chicago: University of Chicago Press, 1983), 200.

154 *"It is, in short":* James, *Principles of Psychology,* 254.

"We ought to say": Edward Reed, *From Soul to Mind* (New Haven, Conn.: Yale University Press, 1997), 208.

155 *"William James taught me":* Richard Poirier, *Poetry and Pragmatism* (Cambridge, Mass.: Harvard University Press, 1992), 92.

"A great deal I owe": Haas, *A Primer,* 34.

"He looked and gasped": Stein, *Autobiography,* 89.

157 *"If there is anything":* Mellow, *Charmed Circle,* 404.

159 *It even found a more elegant:* Howard Gardner, *The Mind's New Science: A History of the Cognitive Revolution* (New York: Basic Books, 1987), 147–55.

The psychologist George Miller: George Miller, "The Magical Number Seven, Plus or Minus Two," *The Psychological Review* 63 (1956): 81–97.

161 *"There are processes of language":* Chomsky's original 1956 paper is accessible here: web.mit.edu/afs/athena.mit.edu/course/6/6.441/www/reading/IT-V2-N3.pdf.

162 *This boundless creativity:* Marc Hauser, Noam Chomsky, Tecumseh Fitch, "The Faculty of Language: What Is It, Who Has It, and How Did It Evolve?" *Science* 298 (2002): 1569–79.

The best evidence: Michal Ben-Shachar, "Neural Correlates of Syntactic Movement: Converging Evidence from Two fMRI Experiments," *Neuroimage* 21 (2004): 1320–36.

163 *Although these Nicaraguan children:* Ann Senghas et al., "Children Creating Core Properties of Language: Evidence from an Emerging Sign Language in Nicaragua," *Science* 305: 1779–82.

"Everybody said the same": Gertrude Stein, *Lectures in America,* 138.

164 *"in enormously long":* Gertrude Stein, *Writings 1932–1946* (New York: Library of America, 1998), 326.

Stein always said: As cited in Meyer, *Irresistible Dictation,* 138.

165 *"When you make a thing":* Stein, *Autobiography,* 28.

"If you keep on doing": Ibid., 230.

Stein later bragged: Ibid., 234.

166 *"Now listen! I'm no fool":* As cited in Mellow, *Charmed Circle,* 404.

167 *"I found out that":* Haas, ed., *A Primer,* 18.

8. Virginia Woolf: The Emergent Self

168 *"I have finally":* Virginia Woolf, *The Diary of Virginia Woolf,* ed. Anne Olivier Bell, 5 vols. (London: Hogarth, 1977–1980), vol. 2, 13.

"Only thoughts and feelings": Virginia Woolf, *Congenial Spirits: The Selected Letters of Virginia Woolf,* ed. Joanne Trautmann Banks (New York: Harvest, 1991), 128.

"They have looked": Virginia Woolf, *The Virginia Woolf Reader* (New York: Harcourt, 1984), 205 [emphasis mine].

"Is it not the task": Ibid., 287.

169 *"was very erratic":* Virginia Woolf, *A Room of One's Own* (New York: Harvest, 1989), 110.

"I press to my centre": Woolf, *The Diary,* vol. 3, 275.

"One must have a whole": As cited in Hermione Lee, *Virginia Woolf* (New York: Vintage, 1996), 407.

"We are the words": Virginia Woolf, *Moments of Being* (London: Pimlico, 2002), 85.

171 *"I intend to keep"*: Woolf, *The Diary*, vol. 5, 64.

"I feel my brains": Lee, *Virginia Woolf*, 187.

"of depositing experience": Woolf, *The Diary*, vol. 4, 231.

172 *"difficult nervous system"*: Ibid., vol. 3, 39.

"It's odd how being ill": Nigel Nicolson and Joanne Trautmann, eds., *The Letters of Virginia Woolf*, 6 vols. (New York: Harcourt Brace Jovanovich, 1975–1980), vol. 3, 388.

"It is no use": Virginia Woolf, *Jacob's Room* (New York: Harvest, 1950), 154.

"At [the age of] forty": Woolf, *The Diary*, vol. 2, 205–06.

"those infinitely obscure": Woolf, *A Room of One's Own*, 89.

"Let us not take it": Virginia Woolf, "Modern Novels," *Times Literary Supplement*, April 10, 1919.

173 *"felt very much"*: Virginia Woolf, *Mrs. Dalloway* (New York: Harvest Books, 1990), 186.

"something central which permeated": Ibid., 31.

"some indescribable outrage": Ibid., 184.

"transcendental theory": Ibid., 151.

"Pointed; dartlike; definite": Ibid., 37.

174 *"world of her own"*: Ibid., 76.

"For there she was": Ibid., 194.

175 *"kind of whole"*: Quentin Bell, *Virginia Woolf: A Biography* (New York: Harvest, 1974), 138.

"The mind receives": Virginia Woolf, *The Common Reader: First Series* (New York: Harvest, 2002), 150.

"There was nobody": Virginia Woolf, *To the Lighthouse* (New York: Harcourt, 1955), 32.

176 *"thought is like"*: Ibid., 33.

"What does one mean": Woolf, *A Room of One's Own*, 97.

"suddenly split off": Ibid.

"like a cloud": Woolf, *The Diary*, vol. 3, 218.

"My hypothesis is the subject": Christopher Butler, *Early Modernism* (Oxford: Oxford University Press, 1994), 92.

"mutations of the self": William James, *The Principles of Psychology* (New York: Dover, 1950), vol. 1, 399.

"The poet has": T. S. Eliot, *The Sacred Wood and Major Early Essays* (New York: Dover, 1998), 32.

177 *"splinters and mosaics"*: Woolf, *The Diary*, vol. 2, 314.

Experiment after experiment: Merlin Donald, *A Mind So Rare* (New York: Norton, 2001), 13–25.

"of multiple channels": Daniel Dennett, *Consciousness Explained* (New York: Back Bay Books, 1991), 253–54.

"Such was the complexity": Woolf, *To the Lighthouse,* 102.

178 *The right hemisphere:* nobelprize.org/medicine/laureates/1981/sperry-lecture.html.

"Everything that we": Stanley Finger, *Minds Behind the Brain* (Oxford: Oxford University Press, 2000), 281.

179 *"Oh, that's easy":* Michael Gazzaniga, *The Social Brain* (New York: Basic Books, 1985), 72.

180 *"Am I here":* Woolf, *The Virginia Woolf Reader,* 253.

"What happens is": Woolf, *The Diary,* vol. 2, 234.

181 *"a thing that you":* Ibid., 171.

"scraps, orts and fragments": Virginia Woolf, *Between the Acts* (New York: Harvest Books, 1970), 189.

"With what magnificent": Virginia Woolf, *The Waves* (New York: Harvest Books, 1950), 261.

"like a light stealing": Woolf, *To the Lighthouse,* 106–07.

182 *"No, ... she did not want":* Ibid., 108.

"But now and again": Ibid., 62.

"Of such moments": Ibid., 105.

"theme, recurring": Virginia Woolf, *The Years* (New York: Harvest, 1969), 369.

183 *These cells can now see:* C. J. McAdams, J.H.R. Maunsell, "Effects of Attention on Orientation Tuning Functions of Single Neurons in Macaque Cortical Area V4," *Journal of Neuroscience* 19 (1999): 431–41.

The illusory self is causing: Steven Yantis, "How Visual Salience Wins the Battle for Awareness," *Nature Neuroscience* 8 (2005): 975–76. And John Reynolds et al., "Attentional Modulation of Visual Processing," *Annual Review of Neuroscience* 27: 611–47.

"our central oyster": Woolf, *The Virginia Woolf Reader,* 248.

184 *While they have no explicit:* Lawrence Weiskrantz, "Some Contributions of Neuropsychology of Vision and Memory to the Problem of Consciousness," in A. Marcel and E. Bisiach, eds., *Consciousness in Contemporary Science* (Oxford: Oxford University Press, 1988).

Brain scans confirm: A. Cowey and P. Stoerig, "The Neurobiology of Blindsight," *Trends in Neuroscience* 14 (1991): 140–45.

Of course, the one thing: Seth Gillihan and Martha Farah, "Is Self Special? A Critical Review of Evidence from Experimental Psychology and Cognitive Neuroscience," *Psychological Bulletin* 131 (2005): 76–97.

185 *"There are no words":* Woolf, *The Waves,* 287.

186 *"What happens":* Christof Koch, *The Quest for Consciousness* (Englewood, Colorado: Roberts and Company, 2004), 271.

"The two percepts": Ibid.

187 *"Life is not"*: Woolf, *The Virginia Woolf Reader*, 287.
"simplify rather than": Virginia Woolf, "Freudian Fiction," *Times Literary Supplement*, March 25, 1920.
188 *"It is quite possible"*: Noam Chomsky, *Language and the Problems of Knowledge* (Cambridge: MIT Press, 1988), 159.
"For it was not knowledge": Woolf, *To the Lighthouse*, 51.
"The final belief": Wallace Stevens, "Adagia," *Opus Posthumous* (New York: Knopf, 1975), 163.
189 *"What she [Lily] wished"*: Woolf, *To the Lighthouse*, 193.
"Instead there are": Ibid., 161.
"the great revelation": Ibid.
"to be on a level": Ibid., 202.
"queer amalgamation": As cited in Julia Briggs, *Virginia Woolf: An Inner Life* (New York: Harcourt, 2005), 210.
"With a sudden intensity": Woolf, *To the Lighthouse*, 208.

Coda

190 *"To say that"*: Richard Rorty, *Contingency, Irony, and Solidarity* (Cambridge: Cambridge University Press, 1989), 8.
191 *"is that all tangible"*: E. O. Wilson, *Consilience* (New York: Vintage, 1999), 291.
192 *"the philosophy of modernism"*: Steven Pinker, *The Blank Slate* (New York: Penguin, 2003), 404.
194 *"it's not clear to him"*: Ian McEwan, *Saturday* (London: Jonathan Cape, 2005), 3.
"It's as if": Ibid.
"what the chances are": Ibid., 128.
"is biological determinism": Ibid., 92.
195 *"the wonder will remain"*: Ibid., 255.
"touched off a yearning": Ibid., 279.
"There's always this": Ibid.
196 *"The greater one's science"*: Vladimir Nabokov, *Strong Opinions* (New York: Vintage, 1990), 44.
197 *"the ability to remain"*: John Keats, *Selected Letters* (Oxford: Oxford University Press, 2002), 41.
"It is imperative": Karl Popper, *Conjectures and Refutations* (New York: Routledge, 2002), 39.

Bibliography

Abbott, Alison. "Music, Maestro, Please!" *Nature* 416 (2002): 12–14.

Ackerman, Diane. *A Natural History of the Senses.* New York: Vintage, 1990.

Acocella, Joan. *The Diary of Vaslav Nijinsky.* Translated by Kyril Fitzlyon. New York: Farrar, Straus and Giroux, 1999.

Alarcon, J. M., et al. "Selective Modulation of Some Forms of Schaffer Collateral-CA1 Synaptic Plasticity in Mice with a Disruption of the CPEB-1 Gene." *Learning and Memory* 11 (2004): 318–27.

Alberini, Christine. "Mechanisms of Memory Stabilization: Are Consolidation and Reconsolidation Similar or Distinct Processes?" *Trends in Neuroscience* 28 (2005).

Altman, Joseph. "Are New Neurons Formed in the Brains of Adult Mammals?" *Science* 135 (1962): 1127–28.

Aquirre, G. K. "The Variability of Human Bold Hemodynamic Responses." *NeuroImage* 8 (1998): 360–69.

Ash, Mitchell G. *Gestalt Psychology in German Culture, 1890–1967: Holism and the Quest for Objectivity.* Cambridge: Cambridge University Press, 1998.

Ashton, Rosemary. *George Eliot: A Life.* New York: Allen Lane, 1996.

———. *G. H. Lewes.* Oxford: Clarendon Press, 1991.

Auerbach, Erich. *Mimesis.* Princeton, N.J.: Princeton University Press, 1974.

Austen, Jane. *Emma.* New York: Modern Library, 1999.

Bailey, C., E. Kandel, and K. Si. "The Persistence of Long-Term Memory." *Neuron* 44 (2004): 49–57.

Balschun, D., et al. "Does cAMP Response Element-Binding Protein Have a Pivotal Role in Hippocampal Synaptic Plasticity and Hippocampus-Dependent Memory?" *Journal of Neuroscience* 23 (2003): 6304–14.

Banfield, Ann. *The Phantom Table.* Cambridge: Cambridge University Press, 2000.

Bao, S., V. Chan, and M. Merzenich. "Cortical Remodeling Induced by Activity of Ventral Tegmental Dopamine Neurons." *Nature* 412 (2001): 79–84.

Bar, M., et al. "Top-Down Facilitation of Visual Recognition." *Proceedings of the National Academy of Sciences* 103 (2006): 449–54.

Barlow, H. B., C. Blakemore, and J. D. Pettigrew. "The Neural Mechanism of Binocular Depth Discrimination." *Journal of Physiology* (London) 193 (1967): 327–42.

Barzun, Jacques. *A Stroll with William James.* Chicago: University of Chicago Press, 1983.

Baudelaire, Charles. *Baudelaire in English.* New York: Penguin, 1998.

———. *Selected Writings on Art and Artists.* Translated by P. E. Charvet. Cambridge: Cambridge University Press, 1972.

———. *Charles Baudelaire: The Mirror of Art.* Translated by Jonathan Mayne. London: Phaidon Press, 1955.

Beare, J. I., ed. *Greek Theories of Elementary Cognition from Alcmaeon to Aristotle.* Oxford: Clarendon Press, 1906.

Beckett, Samuel. *Three Novels.* New York: Grove Press, 1995.

Becks-Malorny, Ulrike. *Cézanne.* London: Taschen, 2001.

Beer, Gillian. *Darwin's Plots.* Cambridge: Cambridge University Press, 2000.

———. *Open Fields.* Oxford: Oxford University Press, 1996.

Bell, Quentin. *Virginia Woolf: A Biography.* New York: Harvest, 1974.

Beluzzi, O., et al. "Becoming a New Neuron in the Adult Olfactory Bulb." *Nature Neuroscience* 6 (2003): 507–18.

Bendor, D., and Q. Wang. "The Neuronal Representation of Pitch in the Primate Auditory Cortex." *Nature* 436 (2005): 1161–65.

Bergson, Henri. *Creative Evolution.* New York: Dover, 1998.

———. *Laughter: An Essay in the Meaning of the Comic.* Los Angeles: Green Integer, 1999.

———. *Time and Free Will.* New York: Harper and Row, 1913.

Berlin, Isaiah. *Three Critics of the Enlightenment.* Princeton, N.J.: Princeton University Press, 2000.

Berman, Paul. "Walt Whitman's Ghost." *The New Yorker,* June 12, 1995, 98–104.

Blackburn, Simon. *Truth: A Guide.* Oxford: Oxford University Press, 2005.

Blakeslee, Sandra. "Cells That Read Minds." *New York Times,* January 10, 2005, sec. D4.

———. "Rewired Ferrets Overturn Theories of Brain Growth." *New York Times,* April 25, 2000, sec. F1.

Boas, Franz. *A Franz Boas Reader.* Chicago: University of Chicago Press, 1989.

Bohan, Ruth. "Isadora Duncan, Whitman, and the Dance." *The Cambridge Companion to Walt Whitman.* Edited by Ezra Greenspan. Cambridge: Cambridge University Press, 1995.

Borges, Jorge Luis. *Collected Ficciones.* New York: Penguin, 1999.

Born, R., and D. Bradley. "Structure and Function of Visual Area MT." *Annual Review of Neuroscience* 28 (2005): 157–89.

Briggs, Julia. *Virginia Woolf: An Inner Life.* New York: Harcourt, 2005.

Brillat-Savarin, Jean Anthelme. *The Physiology of Taste.* Translated by M.F.K. Fisher. New York: Counterpoint Press, 2000.

Browne, Janet. *Charles Darwin: Voyaging.* Princeton, N.J.: Princeton University Press, 1996.

Bucke, Richard Maurice, ed. *Notes and Fragments.* Folcroft, Penn.: Folcroft Library Editions, 1972.

Burrell, Brian. *Postcards from the Brain Museum.* New York: Broadway Books, 2004.

Burrow, J. W. *The Crisis of Reason.* New Haven, Conn.: Yale University Press, 2000.

Butler, Christopher. *Early Modernism.* Oxford: Oxford University Press, 1994.

Caramagno, Thomas. *The Flight of the Mind.* Berkeley: University of California Press, 1992.

Caroll, David, ed. *George Eliot: The Critical Heritage.* London: Routledge and Kegan Paul, 1971.

Carter, William. *Marcel Proust: A Life.* New Haven, Conn.: Yale University Press, 2002.

Cartwright, Nancy. "Do the Laws of Physics State the Facts?" *Pacific Philosophical Quarterly* 61 (1980): 75–84.

Chaudhari, N., et al. "A Novel Metabotropic Receptor Functions as a Taste Receptor." *Nature Neuroscience* 3 (2000): 113–19.

Chip, Herschel B., ed. *Theories of Modern Art: A Source Book by Artists and Critics.* Berkeley: University of California Press, 1984.

Chomsky, Noam. "A Review of B. F. Skinner's 'Verbal Behavior.'" *Language* 35 (1959): 26–58.

———. *Aspects of the Theory of Syntax.* Cambridge: MIT Press, 1965.

———. *The Chomsky Reader.* New York: Pantheon, 1987.

———. *Language and Mind.* New York: Harcourt Brace Jovanovich, 1972.

———. *Language and the Problems of Knowledge.* Cambridge: MIT Press, 1988.

———. *Syntactic Structures.* The Hague: Mouton, 1957.

Chowdury, S. A., and N. Suga. "Reorganization of the Frequency Map of the Auditory Cortex Evoked by Cortical Electrical Stimulation in the Big Brown Bat." *Journal of Neurophysiology* 83 (2000): 1856–63.

Churchland, P. M. "Reduction, Qualia, and the Direct Introspection of Brain States." *Journal of Philosophy* 82 (1985): 8–28.

Coe, C. L., et al. "Prenatal Stress Diminishes Neurogenesis in the Dentate Gyrus of Juvenile Rhesus Monkeys." *Biology of Psychiatry* 10 (2003): 1025–34.

Cohen, Rachel. *A Chance Meeting.* New York: Random House, 2003.

———. "Artist's Model." *The New Yorker,* November 7, 2005, 62–85.

Coleridge, Samuel Taylor. *The Major Works*. Oxford: Oxford University Press, 2000.

Conrad, Peter. *Modern Times, Modern Places*. New York: Knopf, 1999.

Cowey, A., and P. Stoerig. "The Neurobiology of Blindsight." *Trends in Neuroscience* 14 (1991): 140–45.

Craig, A. D. "How Do You Feel? Interoception: The Sense of the Physiological Condition of the Body." *Nature Reviews Neuroscience* 3 (2002), 655–66.

Dalgeish, Tim. "The Emotional Brain." *Nature Reviews Neuroscience* 5 (2004): 582–89.

Damasio, Antonio. *Descartes' Error*. London: Quill, 1995.

———. *The Feeling of What Happens*. New York: Harvest, 1999.

———. *Looking for Spinoza*. London: Vintage, 2003.

Darwin, Charles. *The Autobiography of Charles Darwin*. New York: W. W. Norton, 1993.

———. *The Descent of Man and Selection in Relation to Sex*. New York: Hurst and Co., 1874.

———. *On the Origin of Species by Means of Natural Selection, or the Preservation of Favored Races in the Struggle for Life*. London: John Murray, 1859.

Davidson, Donald. *Essays on Actions and Events*. Oxford: Oxford University Press, 2001.

———. *Inquiries into Truth and Interpretation*. Oxford: Oxford University Press, 2001.

———. *Subjective, Intersubjective, Objective*. Oxford: Oxford University Press, 2001.

Davis, Ronald. "Olfactory Memory Formation in Drosophila: From Molecular to Systems Neuroscience." *Annual Review of Neuroscience* 28 (2005): 275–302.

Dawkins, Richard. *The Selfish Gene*. Oxford: Oxford University Press, 1990.

de Araujo, I. E., et al. "Cognitive Modulation of Olfactory Processing." *Neuron* 46 (2005): 671–79.

Debiec, J., J. LeDoux, and K. Nader. "Cellular and Systems Reconsolidation in the Hippocampus." *Neuron* 36 (2002): 527–38.

Dennett, Daniel. *Consciousness Explained*. New York: Back Bay Books, 1991.

———. *Darwin's Dangerous Idea*. London: Allen Lane, 1995.

———. *Freedom Evolves*. New York: Viking, 2003.

Descartes, René. *Discourse on Method and Meditations of First Philosophy*. Cambridge: Hackett, 1998.

Dewey, John. *Art as Experience*. New York: Perigee, 1934.

———. *Experience and Nature*. New York: Dover, 1958.

———. "Theory of Emotion." *Psychological Review* 1 (1894): 553–69.

Dickinson, Emily. *The Complete Poems of Emily Dickinson*. Boston: Back Bay, 1976.

Dickstein, Morris. *The Revival of Pragmatism*. Chapel Hill, N.C.: Duke University, 1998.

Diggins, John Patrick. *The Promise of Pragmatism*. Chicago: University of Chicago Press, 1994.

Dodd, J. V., et al. "Perceptually Bistable Three-Dimensional Figures Evoke High Choice Probabilities in Cortical Area MT." *Journal of Neuroscience* 21 (2001): 4809–21.

Dodd, Valerie A. *George Eliot: An Intellectual Life*. London: Macmillan, 1990.

Doetsch, Valerie, and Rene Hen. "Young and Excitable: The Function of New Neurons in the Adult Mammalian Brain." *Current Opinion Neurobiology* 15 (2005): 121–28.

Donald, Merlin. *A Mind So Rare*. New York: Norton, 2001.

Doran, Michael, ed. *Conversations with Cézanne*. Berkeley: University of California Press, 2001.

Eco, Umberto. *The Open Work*. London: Hutchinson Radius, 1989.

Edel, Leon. *Henry James: A Life*. London: Flamingo, 1985.

Eliot, George. *Adam Bede*. New York: Penguin Classics, 1980.

———. *Daniel Deronda*. Oxford: Clarendon Press, 1984.

———. *The Lifted Veil: Brother Jacob*. Oxford: Oxford University Press, 1999.

———. *Middlemarch*. London: Norton, 2000.

———. *The Mill on the Floss*. Oxford: Oxford University Press, 1998.

———. "The Natural History of German Life." *The Westminster Review,* July 1856, 28–44.

Eliot, T. S. *The Complete Poems and Plays*. London: Faber and Faber, 1969.

———. *The Sacred Wood and Major Early Essays*. New York: Dover, 1998.

Ellmann, Richard. *James Joyce*. Oxford: Oxford University Press, 1983.

Emerson, Ralph Waldo. *Nature, Addresses, and Lectures*. Boston: Houghton Mifflin, 1890.

———. *Selected Essays, Lectures and Poems*. New York: Bantam, 1990.

Escoffier, Auguste. *Escoffier: The Complete Guide to the Art of Modern Cookery*. New York: Wiley, 1983.

———. *The Escoffier Cookbook: A Guide to the Fine Art of Cookery for Connoisseurs, Chefs, Epicures*. New York: Clarkson Potter, 1941.

Finger, Stanley. *Minds Behind the Brain*. Oxford: Oxford University Press, 2000.

———. *Origins of Neuroscience*. Oxford: Oxford University Press, 1994.

Fish, Stanley. *Is There a Text in This Class?* Cambridge, Mass.: Harvard University Press, 1980.

———. "Professor Sokal's Bad Joke." *New York Times,* May 21, 1996.

Fishman, Y., et al. "Consonance and Dissonance of Musical Chords: Neural Correlates in Auditory Cortex of Monkeys and Humans." *Journal of Neurophysiology* 86 (2001): 2761–88.

Folsom, Ed, and Kenneth M. Price. "Biography." Walt Whitman Archive. http://www.whitmanarchive.org/biography.

Foucault, Michel. *The Order of Things*. New York: Vintage, 1994.

Freud, Sigmund. *On Aphasia.* Translated by E. Stengel. New York: International Universities Press, 1953.

Fry, Roger. *Cézanne: A Study of His Development.* New York: Kessinger, 2004.

Gage, F. H., et al. "Survival and Differentiation of Adult Neural Progenitor Cells Transplanted to the Adult Brain." *Proceedings of the National Academy of Sciences* 92 (1995): 11879–83.

Galison, Peter. *Einstein's Clocks, Poincaré's Maps.* London: Sceptre, 2003.

Galton, Francis. *Hereditary Genius.* London: Macmillan, 1869.

Garafola, Lynn. *Diaghilev's Ballet Russes.* New York: Da Capo, 1989.

Gardner, Howard. *The Mind's New Science: A History of the Cognitive Revolution.* New York: Basic Books, 1987.

Gasquet, Joachim. *Cézanne.* Translated by C. Pemberton. London: Thames and Hudson, 1927.

Gass, William H. *The World Within the Word.* New York: Alfred A. Knopf, 1978.

Gay, Peter. *Freud.* New York: W. W. Norton, 1998.

Gazzaniga, Michael, ed. *The New Cognitive Neurosciences.* 3rd ed. Cambridge: MIT University Press, 2000.

Gazzaniga, Michael, et al. "Some Functional Effects of Sectioning the Cerebral Commissures in Man." *Proceedings of the National Academy of Sciences* 48 (1962): 1765–69.

Geertz, Clifford. "Thick Description: Towards an Interpretative Theory of Culture." In *The interpretation of Cultures.* Edited by C. Geertz. New York: Basic Books, 1973.

Gillihan, Seth, and Martha Farah. "Is Self Special? A Critical Review of Evidence from Experimental Psychology and Cognitive Neuroscience." *Psychological Bulletin* 131 (2005): 76–97.

Goldberg, Elkhonon. *The Executive Brain.* Oxford: Oxford University Press, 2001.

Gopnik, Adam. *Paris to the Moon.* London: Vintage, 2001.

Gould, Elizabeth, et al. "Learning Enhances Adult Neurogenesis in the Hippocampal Formation." *Nature Neuroscience* 2 (1999): 260–65.

Gould, Stephen Jay. "Evolutionary Psychology: An Exchange." *The New York Review of Books* 44 (1997).

———. *The Hedgehog, the Fox, and the Magister's Pox.* New York: Harmony Books, 2003.

———. *The Mismeasure of Man.* New York: W. W. Norton, 1981.

———. *The Structure of Evolutionary Theory.* Cambridge, Mass.: Harvard University Press, 2002.

Greenberg, Clement. "Modernist Painting." *Art and Literature* 4 (1965): 193–201.

Gross, C. G. "Genealogy of the "Grandmother Cell." *Neuroscientist* 8 (2002): 512–18.

———. "Neurogenesis in the Adult Brain: Death of a Dogma." *Nature Reviews Neuroscience* 1 (2000): 67–72.

Haas, Robert Bartlett, ed. *A Primer for the Gradual Understanding of Gertrude Stein.* Los Angeles: Black Sparrow, 1971.

Hacking, I. *The Taming of Chance.* Cambridge: Cambridge University Press, 1990.

———. "Wittgenstein the Psychologist." *New York Review of Books* (1982).

Haight, Gordon, ed. *George Eliot's Letters.* New Haven, Conn.: Yale University Press, 1954–1978.

Hauser, Marc, Noam Chomsky, and Tecumseh Fitch. "The Faculty of Language: What Is It, Who Has It, and How Did It Evolve?" *Science* 298 (2002): 1569–79.

Heisenberg, Werner. *Philosophic Problems of Nuclear Science.* New York: Pantheon, 1952.

Herz, Rachel. "The Effect of Verbal Context on Olfactory Perception." *Journal of Experimental Psychology: General* 132 (2003): 595–606.

Hill, Peter. *Stravinsky: The Rite of Spring.* Cambridge: Cambridge University Press, 2000.

Hollerman, J. R., and W. Schultz. "Dopamine Neurons Report an Error in the Temporal Prediction of Reward During Learning." *Nature Neuroscience* 1 (1998): 304–09.

Holmes, Richard. *Coleridge: Darker Reflections.* London: HarperCollins, 1998.

Hoog, Michel. *Cézanne.* London: Thames and Hudson, 1989.

Horgan, John. *The End of Science.* London: Abacus, 1996.

Hubel, D. H., T. N. Weisel, and S. LeVay. "Plasticity of Ocular Dominance Columns in Monkey Striate Cortex." *Philosophical Transactions of the Royal Society of London Biology Letters* 278 (1977): 377–409.

Husserl, Edmund. *General Introduction to Phenomenology.* New York: Allen and Unwin, 1931.

Huxley, Aldous. *Literature and Science.* New Haven, Conn.: Leete's Island Books, 1963.

Huxley, Thomas. "On the Hypothesis That Animals Are Automata, and Its History." *Fortnightly Review* (1874): 575–77.

Jackendoff, Ray. *Patterns in the Mind.* New York: Basic Books, 1994.

Jackendoff, Ray, and Steven Pinker. "The Nature of the Language Faculty and Its Implication for Evolution of Language." *Cognition* 97 (2005): 211–25.

James, Henry. *The Art of Criticism.* Chicago: University of Chicago Press, 1986.

———. *The Figure in the Carpet and Other Stories.* London: Penguin, 1986.

James, Kenneth. *Escoffier: The King of Chefs.* London: Hambledon and London, 2002.

James, William. "The Consciousness of Lost Limbs." *Proceedings of the American Society for Psychical Research* 1 (1887): 249–58.

———. *Pragmatism.* New York: Dover, 1995.

———. *The Principles of Psychology.* Vol. 1. New York: Dover, 1950.

———. *The Principles of Psychology.* Vol. 2. New York: Dover, 1950.

———. "What Is an Emotion?" *Mind* 9 (1884): 188–205.

———. *Writings: 1878–1899.* New York: Library of America, 1987.

———. *Writings: 1902–1910.* New York: Library of America, 1987.

———. *The Varieties of Religious Experience.* New York: Penguin Classics, 1982.

Jarrell, Randall. *No Other Book.* New York: HarperCollins, 1999.

Joyce, James. *Ulysses.* New York: Vintage, 1990.

Kandel, Eric. *In Search of Memory.* New York: Norton, 2006.

Kandel, Eric, James Schwartz, and Thomas Jessell. *Principles of Neural Science.* 4th ed. New York: McGraw Hill, 2000.

Kant, Immanuel. *The Critique of Pure Reason.* Translated by J.M.D. Meiklejohn. New York: Prometheus Books, 1990.

Kaplan, M. S. "Neurogenesis in the Three-Month-Old Rat Visual Cortex." *Journal of Comparative Neurology* 195 (1981): 323–38.

Kaplan, Michael, and Ellen Kaplan. *Chances Are . . .* New York: Viking, 2006.

Kaufmann, Michael Edward. "Gertrude Stein's Re-Vision of Language and Print in Tender Buttons." *Journal of Modern Literature* 15 (1989): 447–60.

Keats, John. *Selected Letters.* Oxford: Oxford University Press, 2002.

Kelly, Thomas. *First Nights.* New Haven, Conn.: Yale University Press, 2000.

Kermode, Frank. *History and Value.* Oxford: Clarendon Press, 1988.

Keynes, R. D., ed. *Charles Darwin's Beagle Diary.* Cambridge: Cambridge University Press, 2001.

Kitcher, Philip. *In Mendel's Mirror.* Oxford: Oxford University Press, 2003.

Koch, Christof. *The Quest for Consciousness.* Englewood, Colo.: Roberts and Company, 2004.

Kolocotroni, Vassiliki, Jane Goldman, and Olga Taxidou, eds. *Modernism: An Anthology of Sources and Documents.* Chicago: University of Chicago Press, 1998.

Kozorovitskiy, Y., et al. "Experience Induces Structural and Biochemical Changes in the Adult Primate Brain." *Proceedings of the National Academy of Sciences* 102 (2005): 17478–82.

Kuhn, Thomas. *The Structure of Scientific Revolutions.* 3rd ed. Chicago: University of Chicago Press, 1996.

Kummings, Donald, and J. R. LeMaster, eds. *Walt Whitman: An Encyclopedia.* New York: Garland, 1998.

Landy, Joshua. *Philosophy as Fiction: Self, Deception, and Knowledge in Proust.* Oxford: Oxford University Press, 2004.

Laurent, G. "Odor Encoding as an Active, Dynamical Process." *Annual Review of Neuroscience* 24 (2001): 263–97.

Lee, Hermione. *Virginia Woolf.* New York: Vintage, 1996.

Levine, George, ed. *Cambridge Companion to George Eliot.* Cambridge: Cambridge University Press, 2001.

Lewes, George Henry. *Comte's Philosophy of Science.* London, 1853.

———. *The Life of Goethe.* 2nd ed. London: Smith, Elder, and Co., 1864.

———. *The Physical Basis of Mind. With Illustrations. Being the Second Series of Problems of Life and Mind.* London: Trübner, 1877.

Lewontin, Richard. *Biology as Ideology.* New York: Harper Perennial, 1993.

Lewontin, Richard, and Stephen Jay Gould. "The Spandrels of San Marco and the Panglossian Paradigm: A Critique of the Adaptationist Programme." *Proceedings of the Royal Society of London* 205, no. 1161 (1979): 581–98.

Lindemann, B., et al. "The Discovery of Umami." *Chemical Senses* 27 (2002): 843–44.

Litvin, O., and K. V. Anokhin. "Mechanisms of Memory Reorganization During Retrieval of Acquired Behavioral Experience in Chicks: The Effects of Protein Synthesis Inhibition in the Brain." *Neuroscience and Behavioral Physiology* 30 (2000): 671–78.

Livingstone, Margaret. *Vision and Art: The Biology of Seeing.* New York: Harry Abrams, 2002.

Loving, Jerome. *Walt Whitman.* Berkeley: University of California Press, 1999.

Luria, A. R. *The Mind of a Mnemonist.* Cambridge, Mass.: Harvard University Press, 1995.

Ma, X., and N. Suga. "Augmentation of Plasticity of the Central Auditory System by the Basal Forebrain and/or Somatosensory Cortex." *Journal of Neurophysiology* 89 (2003): 90–103.

———. "Long-Term Cortical Plasticity Evoked by Electrical Stimulation and Acetylcholine Applied to the Auditory Cortex." *Proceedings of the National Academy of Sciences,* June 16, 2005.

Mahon, Basil. *The Man Who Changed Everything: The Life of James Clerk Maxwell.* London: John Wiley, 2003.

Maia, T., and J. McClelland. "A Reexamination of the Evidence for the Somatic Marker Hypothesis: What Participants Really Know in the Iowa Gambling Task." *Proceedings of the National Academy of Sciences* 101 (2004): 16075–80.

Mainland, J. D., et al. "One Nostril Knows What the Other Learns." *Nature* 419 (2002): 802.

Malcolm, Janet. "Someone Says Yes to It." *The New Yorker,* June 13, 2005, 148–65.

Martin, Kelsey, et al. "Synapse-Specific, Long-Term Facilitation of Aplysia Sensory to Motor Synapses: A Function for Local Protein Synthesis in Memory Storage." *Cell* 91 (1997): 927–38.

McAdams, C. J., and J.H.R. Maunsell. "Effects of Attention on Orientation-Tuning Functions of Single Neurons in Macaque Cortical Area V4." *Journal of Neuroscience* 19 (1999): 431–41.

McEwan, Ian. *Saturday.* London: Jonathan Cape, 2004.

McGee, Harold. *On Food and Cooking.* New York: Scribner, 2004.

McGinn, Colin. *The Mysterious Flame.* New York: Basic Books, 1999.

McNeillie, Andrew, and Anne Olivier Bell, eds. *The Diary of Virginia Woolf.* 5 vols. New York: Harcourt Brace Jovanovich, 1976–1984.

Mellow, James. *Charmed Circle: Gertrude Stein and Company.* London: Phaidon, 1974.

Melville, Herman. *Redburn, White-Jacket, Moby Dick.* New York: Library of America, 1983.

Menand, Louis. *The Metaphysical Club.* New York: Farrar, Straus and Giroux, 2001.

———. ed. *Pragmatism: A Reader.* New York: Vintage, 1997.

Meyer, Leonard. *Emotion and Meaning in Music.* Chicago: University of Chicago Press, 1961.

———. *Explaining Music.* Berkeley: University of California Press, 1973.

———. *Music, the Arts, and Ideas.* Chicago: University of Chicago Press, 1994.

———. *The Spheres of Music.* Chicago: University of Chicago Press, 2000.

Meyer, Steven. *Irresistible Dictation.* Palo Alto: Stanford University Press, 2001.

———. "The Physiognomy of the Thing: Sentences and Paragraphs in Stein and Wittgenstein." *Modernism/Modernity* 5.1 (1998): 99–116.

Milekic, M. H., and C. M. Alberini. "Temporally Graded Requirement for Protein Synthesis Following Memory Reactivation." *Neuron* 36 (2002): 521–25.

Miller, Edwin Haviland, ed. *Walt Whitman: The Correspondence.* 6 vols. New York: New York University Press, 1961–1977.

Miller, George. "The Magical Number Seven, Plus or Minus Two." *The Psychological Review* 63 (1956): 81–97.

Mitchell, Silas Weir. *Injuries of Nerves, and Their Consequences.* Philadelphia: Lippincott, 1872.

Muotri, A. R., et al. "Somatic Mosaicism in Neuronal Precursor Cells Mediated by L1 Retrotransposition." *Nature* 435 (2005): 903–10.

Myers, Gerald E. *William James: His Life and Thought.* New Haven, Conn.: Yale University Press, 1986.

Nabokov, Vladimir. *Lectures on Literature.* New York: Harcourt Brace, 1980.

———. *Strong Opinions.* New York: Vintage, 1990.

Nader, K., et al. "Characterization of Fear Memory Reconsolidation." *Journal of Neuroscience* 24 (2004): 9269–75.

Nader, K., et al. "Fear Memories Require Protein Synthesis in the Amygdala for Reconsolidation after Retrieval." *Nature* 406: 686–87.

Nelson, G., et al. "An Amino-Acid Taste Receptor." *Nature* 416 (2002): 199–202.

Nicolson, Nigel, and Trautmann, Joanne, eds., *The Letters of Virginia Woolf.* 6 vols. New York: Harcourt Brace Jovanovich, 1975–1980.

Nietzsche, Friedrich. *The Gay Science.* Translated by Walter Kaufmann. New York: Vintage, 1974.

———. *The Genealogy of Morals.* Translated by Walter Kaufmann. New York: Vintage, 1989.

Nottebohm, Fernando. "Neuronal Replacement in Adulthood." *Annals of the New York Academy of Science* 457 (1985): 143–61.

Olby, Robert. *The Path to the Double Helix.* London: Macmillan, 1974.

Otis, Laura, ed. *Literature and Science in the Nineteenth Century.* Oxford: Oxford University Press, 2002.

Papassotiropoulos, A., et al. "The Prion Gene Is Associated with Human Long-Term Memory." *Human Molecular Genetics* 14 (2005): 2241–46.

Patel, A., and E. Balaban. "Temporal Patterns of Human Cortical Activity Reflect Tone Sequence Structure." *Nature* 404 (2002): 80–83.

Pater, William. *The Renaissance: Studies in Art and Poetry.* Oxford: Oxford University Press, 1998.

Peirce, Charles Sanders. *Peirce on Signs: Writings on Semiotics.* Chapel Hill: University of North Carolina Press, 1991.

Peretz, I., and R. Zatorre. "Brain Organization for Music Processing," *Annual Review of Psychology* 56 (2005): 89–114.

Peretz, I., and R. Zatorre, eds. *The Cognitive Neuroscience of Music.* Oxford: Oxford University Press, 2003.

Perez-Gonzalez, D., et al. "Novelty Detector Neurons in the Mammalian Auditory Midbrain." *European Journal of Neuroscience* 11 (2005): 2879–85.

Perry, Ralph Barton. *The Thought and Character of William James.* 2 vols. Boston: Little, Brown, 1935.

Pincock, Stephen. "All in Good Taste." *FT Magazine,* June 25, 2005, 13.

Pinker, Steven. *The Blank Slate.* New York: Penguin, 2003.

———. *The Language Instinct.* London: Penguin, 1994.

———. *How the Mind Works.* New York: W. W. Norton, 1999.

———. *Words and Rules.* New York: Basic Books, 2000.

Plato. *Timaeus.* Translated by Donald Zeyl. New York: Hackett, 2000.

Poirier, Richard. *Poetry and Pragmatism.* Cambridge, Mass.: Harvard University Press, 1992.

———. *Trying It Out in America.* New York: Farrar, Straus and Giroux, 1999.

Polley, D. B., et al. "Perceptual Learning Directs Auditory Cortical Map Reorganization Through Top-Down Influences." *Journal of Neuroscience* 26 (2006): 4970–82.

Popper, Karl. *Conjectures and Refutations.* New York: Routledge, 2002.

———. *Objective Knowledge.* Oxford: Oxford University Press, 1972.

Proust, Marcel. *The Captive and the Fugitive.* Vol. V. New York: Modern Library, 1999.

———. *The Guermantes Way.* Vol III. New York: Modern Library, 1998.

———. *Letters to His Mother.* New York: Greenwood Press, 1973.

———. *On Art and Literature.* New York: Carrol and Graf, 1997.

———. *Pleasures and Regrets.* London: Peter Owen, 1986.

———. *Sodom and Gomorrah.* Vol IV. New York: Modern Library, 1998.

———. *Swann's Way.* Vol. I. New York: Modern Library, 1998.

———. *Time Regained.* Vol. VI. New York: Modern Library, 1999.

———. *Within a Budding Grove.* Vol. II. New York: Modern Library, 1998.

Quine, V. W. *From a Logical Point of View.* Cambridge, Mass.: Harvard University Press, 2003.

———. *Ontological Relativity and Other Essays.* New York: Columbia University Press, 1977.

———. *Quintessence.* Cambridge, Mass.: Belknap Press, 2004.

Quiroga, R., et al. "Invariant Visual Representation by Single Neurons in the Human Brain." *Nature* 435 (2005): 1102–07.

Rakic, P. "Limits of Neurogenesis in Primates." *Science* 227 (1985): 1054–56.

Ramon y Cajal, Santiago. *Advice for a Young Investigator.* Cambridge: MIT Press, 2004.

———. *Nobel Lectures, Physiology or Medicine, 1901–1921.* Amsterdam: Elsevier Publishing, 1967.

Reed, Edward. *From Soul to Mind.* New Haven, Conn.: Yale University Press, 1997.

Renton, Alex. "Fancy a Chinese?" *Observer Food Magazine,* July 2005, 27–32.

Rewald, John. *Cézanne.* New York: Harry Abrams, 1986.

Reynolds, John, et al. "Attentional Modulation of Visual Processing." *Annual Review of Neuroscience* 27: 611–47.

Richardson, Alan. *British Romanticism and the Science of the Mind.* Cambridge: Cambridge University Press, 2001.

Richardson, Richard. *Emerson: The Mind on Fire.* Berkeley: University of California Press, 1996.

Richter, Joel. "Think Globally, Translate Locally: What Mitotic Spindles and Neuronal Synapses Have in Common." *Proceedings of the National Academy of Sciences* 98 (2001): 7069–71.

Rilke, Rainer Maria. *Letters on Cézanne.* London: Vintage, 1985.

Rizzolatti, Giacomo, Leonardo Fogassi, and Vittorio Gallese. "Neurophysiological Mechanisms Underlying the Understanding of Imitation and Action." *Nature Reviews Neuroscience* 1 (2001): 661–70.

Rorty, Richard. *Contingency, Irony, and Solidarity.* Cambridge: Cambridge University Press, 1989.

———. *Essays on Heidegger and Others.* Cambridge: Cambridge University Press, 1991.

———. *Objectivity, Relativism, and Truth.* Cambridge: Cambridge University Press, 1991.

———. *Philosophy and the Mirror of Nature.* Oxford: Basil Blackwell, 1978.

———. *Philosophy and Social Hope.* New York: Penguin, 1999.

Rose, Steven. *Lifelines: Biology, Freedom, Determinism.* London: Allen Lane, 1997.

———. *The 21st Century Brain.* London: Jonathan Cape, 2005.

Rosen, Charles. *Arnold Schoenberg.* Chicago: University of Chicago Press, 1996.

Ross, Alex. "Prince Igor." *The New Yorker,* November 6, 2000.

———. "Whistling in the Dark: Schoenberg's Unfinished Revolution." *The New Yorker,* February 18, 2002.

Ryan, Judith. *The Vanishing Subject.* Chicago: University of Chicago Press, 1991.

Sacks, Oliver. *An Anthropologist on Mars*. London: Picador, 1995.

——. *The Man Who Mistook His Wife for a Hat*. London: Picador, 1985.

——. *Seeing Voices*. New York: Vintage, 2000.

Sanfey, Alan, and and Jonathan Cohen. "Is Knowing Always Feeling?" *Proceedings of the National Academy of Sciences* 101 (2004): 16709–10.

Santarelli, Luca, et al. "Requirement of Hippocampal Neurogenesis for the Behavioral Effects of Antidepressants." *Science* 301 (2003): 805–08.

Schjeldahl, Peter. "Two Views." *The New Yorker*, July 11, 2005.

Schmidt-Hieber, C. "Enhanced Synaptic Plasticity in Newly Generated Granule Cells of the Adult Hippocampus." *Nature* 429 (2004): 184–87.

Schoenberg, Arnold. *Style and Idea: Selected Writings*. London: Faber, 1975.

——. *Theory of Harmony*. Berkeley: University of California Press, 1983.

Schoenfeld, M., et al. "Functional MRI Tomography Correlates of Taste Perception in the Human Primary Taste Cortex." *Neuroscience* 127 (2004): 347–53.

Schultz, Wolfram, et al. "Neuronal Coding of Prediction Errors." *Annual Review of Neuroscience* 23: 473–500.

Senghas, Ann, et al. "Children Creating Core Properties of Language: Evidence from an Emerging Sign Language in Nicaragua." *Science* 305: 1779–82.

Shapin, Steven. *The Scientific Revolution*. Chicago: University of Chicago Press, 1996.

Sharma, J., A. Angelucci, and M. Sur. "Induction of Visual Orientation Modules in Auditory Cortex." *Nature* 404 (2000): 841–47.

Shattuck, Roger. *Proust's Way*. New York: W. W. Norton, 2001.

Shuttleworth, Sally. *George Eliot and 19th Century Science*. Cambridge: Cambridge University Press, 1984.

Si, K., et al. "A Neuronal Isoform of CPEB Regulates Local Protein Synthesis and Stabilizes Synapse-Specific Long-Term Facilitation in Aplysia." *Cell* 115 (2003): 893–904.

Si, K., E. Kandel, and S. Lindquist. "A Neuronal Isoform of the Aplysia CPEB Has Prion-Like Properties." *Cell* 115 (2003): 879–91.

Sifton, Sam. "The Cheat." *New York Times Magazine*, May 8, 2005.

Silvers, Robert. *Hidden Histories of Science*. New York: New York Review of Books Press, 1995.

Simms, Byran. *The Atonal Music of Arnold Schoenberg*. Oxford: Oxford University Press, 2000.

Sinclair, May. *Mary Oliver: A Life*. New York: New York Review of Books Press, 2002.

Specter, Michael. "Rethinking the Brain." *The New Yorker*, July 23, 2001, 42–65.

Sperry, Roger. "Cerebral Organization and Behavior." *Science* 133 (1961): 1749–57.

——. "Some Effects of Disconnecting the Cerebral Hemispheres." Nobel lecture, 1981. Available from nobelprize.org/medicine/laureates/1981/sperry-lecture.html (accessed March 5, 2005).

Squire, Larry, and Eric Kandel. *Memory: From Mind to Molecules.* New York: Owl Books, 1999.

Stein, Gertrude. *As Fine as Melanctha.* New Haven, Conn.: Yale University Press, 1954.

———. *The Autobiography of Alice B. Toklas.* London: Penguin Classics, 2001.

———. *Everybody's Autobiography.* Cambridge: Exact Change, 1993.

———. *Lectures in America.* Boston: Beacon Press, 1985.

———. *Picasso.* Boston: Beacon Press, 1959.

———. *The Selected Writings of Gertrude Stein.* New York: Vintage, 1990.

———. *Writings: 1903–1932.* New York: Library of America, 1998.

———. *Writings 1932–1946.* New York: Library of America, 1998.

Steingarten, Jeffrey. *It Must've Been Something I Ate.* New York: Vintage, 2003.

Stevens, Wallace. *The Necessary Angel.* New York: Vintage, 1951.

———. *Opus Posthumous.* New York: Knopf, 1975.

———. *The Palm at the End of the Mind.* New York: Vintage, 1967.

Stoddard, Tim. "Scents and Sensibility." *Columbia,* spring 2005, 17–21.

Stravinsky, Igor. *Poetics of Music.* New York: Vintage, 1947.

———. *Chronicle of My Life.* Gollancz: London, 1936.

Stravinsky, Igor, and Robert Craft. *Conversations with Igor Stravinsky.* London: Faber, 1979.

———. *Memories and Commentaries.* London: Faber and Faber, 1960.

———. *Expositions and Development.* London: Faber and Faber, 1962.

Stravinsky, Vera, and Robert Craft, eds. *Stravinsky in Pictures and Documents.* New York: Simon and Schuster, 1978.

Suga, N., and E. Gao. "Experience-Dependent Plasticity in the Auditory Cortex and the Inferior Colliculus of Bats: Role of the Cortico-Fugal System." *Proceedings of the National Academy of Sciences* 5 (2000): 8081–86.

Suga, N., et al. "The Corticofugal System for Hearing: Recent Progress." *Proceedings of the National Academy of Sciences* 22 (2000): 11807–14.

Sur, M., and J. L. Rubenstein. "Patterning and Plasticity of the Cerebral Cortex." *Science* 310 (2005): 805–10.

Sylvester, David. *About Modern Art.* London: Pimlico, 1996.

Tadie, Jean Yves. *Marcel Proust: A Life.* New York: Penguin, 2000.

Tanaka, S. "Dopaminergic Control of Working Memory and Its Relevance to Schizophrenia: A Circuit Dynamics Perspective." *Neuroscience* 139 (2005): 153–71.

Taruskin, Richard. *Stravinsky and the Russian Tradition.* Oxford: Oxford University Press, 1996.

Toorn, Pieter van den. *Stravinsky and the Rite of Spring.* Oxford: Oxford University Press, 1987.

Tramo, M., et al. "Neurobiological Foundations for the Theory of Harmony in Western Tonal Music." *Annals of the New York Academy of Science* 930 (2001): 92–116.

Traubel, Horace. *Intimate with Walt: Selections from Whitman's Conversations with Horace Traubel, 1882–1892.* Des Moines: University of Iowa Press, 2001.

Trubek, Amy. *Haute Cuisine: How the French Invented the Culinary Profession.* Philadelphia: University of Pennsylvania Press, 2001.

Updike, John. *Self-Consciousness.* New York: Fawcett, 1990.

Vollard, Ambroise. *Cézanne.* New York: Dover, 1984.

Wade, Nicholas. "Explaining Differences in Twins." *New York Times,* July 5, 2005.

Walsh, Stephen. *Igor Stravinsky: A Creative Spring.* Berkeley: University of California Press, 2002.

Wang, L., et al. "Evidence for Peripheral Plasticity in Human Odour Response." *Journal of Physiology* (January 2004): 236–44.

Weiner, Jonathan. *Time, Love, Memory.* New York: Knopf, 1999.

Weiskrantz, Lawrence. "Some Contributions of Neuropsychology of Vision and Memory to the Problem of Consciousness." In *Consciousness in Contemporary Science,* edited by A. Marceland E. Bisiach. Oxford: Oxford University Press, 1988.

West-Eberhard, Mary Jane. *Developmental Plasticity and Evolution.* Oxford: Oxford University Press, 2003.

Whitman, Walt. *Democratic Vistas and Other Papers.* Amsterdam: Fredonia Books, 2002.

——. *Leaves of Grass: The "Death-Bed" Edition.* New York: Random House, 1993.

Wilshire, Bruce, ed. *William James: The Essential Writings.* Albany: State University of New York, 1984.

Wilson, Edmund. *Axel's Castle.* New York: Charles Scribner's Sons, 1931.

Wilson, E. O. *Consilience: The Unity of Knowledge.* New York: Vintage, 1999.

——. *Sociobiology: The New Synthesis.* Cambridge, Mass.: Belknap Press, 2000.

Woolf, Virginia. *Between the Acts.* New York: Harvest Books, 1970.

——. *Collected Essays.* London: Hogarth Press, 1966–1967.

——. *The Common Reader: First Series.* New York: Harvest, 2002.

——. *Jacob's Room.* New York: Harvest, 1950.

——. *Moments of Being.* London: Pimlico, 2002.

——. *Mrs. Dalloway.* New York: Harvest Books, 1990.

——. *A Room of One's Own.* New York: Harvest, 1989.

——. *To the Lighthouse.* New York: Harcourt, 1955.

——. *The Virginia Woolf Reader.* New York: Harcourt, 1984.

——. *The Waves.* New York: Harvest Books, 1950.

——. *The Years.* New York: Harvest, 1969.

——. *Congenial Spirits: The Selected Letters of Virginia Woolf.* Edited by Joanne Trautmann Banks. New York: Harvest, 1991.

——. *The Diary of Virginia Woolf.* Edited by Anne Olivier Bell. 5 vols. London: Hogarth, 1977–1980.

Index

Coming in February 2009

HOW WE DECIDE

A REVEALING LOOK AT THE NEW SCIENCE OF DECISION-MAKING — AND HOW IT CAN HELP US MAKE BETTER CHOICES

Since Plato, philosophers have described the decision-making process as either rational or emotional: we carefully deliberate or we "blink" and go with our gut. But as scientists break open the mind's black box with the latest tools of neuroscience, they're discovering that this is not how the mind works. Our best decisions are a finely tuned blend of both feeling and reason — and the precise mix depends on the situation. In *How We Decide,* Jonah Lehrer sets out to answer two questions that are of interest to just about anyone, from CEOs to firefighters: How does the human mind make decisions? And how can we make those decisions better?

ISBN: 978-0-618-62011-1